VGM Opportunities Se

OPPORTUNITIES IN **FARMING AND AGRICULTURE CAREERS**

William C. White
Donald N. Collins

Revised by
Adrian A. Paradis

Foreword by
C. Coleman Harris
Executive Secretary
National FFA Organization
Future Farmers of America

VGM Career Horizons
a division of *NTC Publishing Group*
Lincolnwood, Illinois USA

Cover Photo Credits
All photos courtesy of the U.S. Department of Agriculture.

Library of Congress Cataloging-in-Publication Data
White, William C.
Opportunities in farming and agriculture careers / William C. White, Donald N. Collins ; revised by Adrian A. Paradis ; foreword by C. Coleman Harris.
p. cm. — (VGM opportunities series)
Rev. ed. of: Opportunities in agricultural careers, c1988.
Includes bibliographical references (p.).
ISBN 0-8442-4580-1 (h). — ISBN 0-8442-4582-8 (p)
1. Agriculture—Vocational guidance—United States.
2. Agricultural industries—Vocational guidance—United States.
I. Collins, Donald N. II. Paradis, Adrian A. III. White, William C. Opportunities in agriculture careers. IV. Title. V. Series.
S494.5.A4W48 1995
338.1'023—dc20 95-21860
CIP

Published by VGM Career Horizons, a division of NTC Publishing Group
4255 West Touhy Avenue
Lincolnwood (Chicago), Illinois 60646-1975, U.S.A.

Manufactured in the United States of America.

5 6 7 8 9 VP 9 8 7 6 5 4 3 2 1

CONTENTS

ABOUT THE AUTHORS

William C. White is an agronomist with professional experience at land-grant universities and with private industry. His education at Virginia Polytechnic Institute and Iowa State University was in agronomy and soil fertility, and he earned his Ph. D. degree in soil fertility at Iowa State. While in Iowa, his research was some of the earliest to identify the nature and extent of nitrogen carryover in soils from commercial fertilizer.

Much of his career has centered on transferring information from research and technical fields related to fertilizers to users, and on producing information showing the extent and relevance of fertilizer use to food and fiber production. Contributions in his career range from organizing intensive soil testing programs to working on world supply/demand balances for fertilizers at the Food and Agricultural Organization of the United Nations in Rome. His career with fertilizer-related services has spanned a period unequal to any other in terms of humankind using chemistry and commerce to expand fertilizer use as a means of improving the fertility and productivity level of soils worldwide.

He is a fellow of the American Society of Agronomy, the Soil Science Society of America, and the Crop Science Society of America.

Donald N. Collins has been involved in communication and information activities since graduating from the University of Missouri with a degree in agricultural journalism and graduate study. His career has included service as assistant professor for the university and in commercial radio. For a number of years, he served as head of communications, information, and public relations for the nation's commercial fertilizer industry association before joining Mr. White in forming Resource Washington, Inc., in the nation's capital.

He has received distinguished service recognition from the Agricultural Communicators in Education, the American Agricultural Editors' Association, and the Agricultural Relations Council. He maintains close working relationships with members of the National Association of Farm Broadcasters, as well as with the general news media.

FOREWORD

Modern agriculture offers a future of rapidly expanding career opportunities, especially in the twin areas of technology and business. Some of these will be in production, whether as a citrus grower in Florida or as a broiler producer in Georgia. Many other opportunities will emerge from America's leading-edge fields of biotechnology and communications, perhaps as a genetic researcher in California or as a market analyst in New York. Still more agri-careers, nonexistent today, will develop in food processing and technology, and in nonfood uses of agricultural crops.

The diverse opportunities in agricultural careers cover more than 200 career fields available to those interested in producing, processing, and marketing food and fiber—or who wish to serve this nation's largest industry, which employs one-fifth of the nation's work force.

The number of career opportunities is much greater than two decades ago, and it is still growing. The primary factors driving this growth include the extensive application of technology and efficiency in agriculture, the expansion of scientific knowledge and research in producing and processing farm products, and the rapid advances in methods of communication and marketing.

Many of today's agricultural careers are enmeshed in the biological sciences and will be even more so in the twenty-first century. Those concerned with the growing of crops and livestock, and with related processing and distribution, will face the pressing challenge of achieving razor-sharp business management capabilities.

Careers in agriculture offer a unique combination of fulfillment in life processes—and telling about them—as well as in the world of business management. Additionally, modern agriculture offers exciting opportunities in information technology and education. All sectors are communication and learning intensive. Classroom back-grounding, satellite information, near instantaneous commodity market data, and operational cost analyses by computers are only several examples of the insatiable diet of information demanded by those pursuing, and holding, careers in agriculture.

It has been estimated that more than 48,000 career opportunities in agriculture are available annually for college graduates. Although salary levels vary, most are in line with similar positions in other industries. And, while a rural background is often helpful, it is by no means essential in many careers.

The foundation for these careers often is laid at the intermediate or high school level with the vocational agriculture instructor (FFA advisor), 4-H Club leader, biology or other science teacher, or with a career counselor. These knowledgeable advisors have a critical role in providing sound information and counsel to those inquiring about agricultural careers. It is with this thought in mind that this book was prepared.

Agriculture truly is a "growth" industry. Worldwide, more than five billion people depend on the production of food and fiber for their well-being. This number is projected to continue to rise; and, as it does, so will the opportunities for those who seek satisfying, rewarding careers in agriculture.

C. Coleman Harris
Executive Secretary
National FFA Organization
Future Farmers of America

INTRODUCTION

There are career opportunities in farming and agriculture for all interests, ages, and levels of education. Don't be fooled by the false perception that farming and agriculture offer dead-end or dreary jobs. Nothing is further from the truth. Read ahead and learn the facts for yourself.

In this revised edition, the editors felt it advisable to expand the text to include information on support and farming careers. Support careers include those innumerable office and professional positions found in business and industry which are essential to most companies. Because farming is in a transitional stage, it is important that readers interested in this field as a possible career choice know the current and anticipated opportunities which are available today, as well as the pros and cons of owning and/or working on a farm.

Before you begin, a brief explanation of terms. Although the meaning of the word *agriculture* includes farm operations, *farming,* which is discussed in Chapter 3, refers to occupations necessary in the actual operation of any farm or ranch. On the other hand, *agriculture* is the umbrella term used in this book for all the non-farm activities and job opportunities covered in the subsequent chapters. As you will see, farming and agriculture positions are as different as A is from Z, but the careers described in the agricultural chapters (Chapters 4–10) all depend on the production of the farmer and rancher!

CHAPTER 1

AGRICULTURE—A UNIVERSAL INDUSTRY

Early one morning, long before the rest of us were up, a dairy farmer on a northern Maine farm was milking his forty Holstein cows. A little later the milk truck containing the previous day's fluid was pumping its load into a processing plant outside Boston, where some would be pasteurized and bottled while the rest would be converted to other dairy products. By 9:05 A.M. in Manhattan's World Trade Center, brokers were exchanging bid and ask prices for unseen bales of cotton in a Memphis warehouse, as well as for five boxcars of eggs on a siding in Nebraska.

Out in Ohio and Illinois, farmers were plowing corn and soybeans, while further west in St. Louis, scientists were peering into microscopes hoping to isolate genes that might help bolster wheat production. In another laboratory a young woman was studying the larvae of a corn borer.

In the Kansas City stockyards, beef cattle were noisily shuffling off the cars into the pen, soon to be slaughtered and converted to steaks and other animal products. Up north on the Canadian flatlands, huge combines started their day's work of harvesting acres of wheat destined for eastern flour mills. West of the Great Plains foresters were marking timber to be harvested and converted at nearby sawmills into lumber to be shipped to a Vancouver housing project, while higher in the Rockies, soil conservationists studied the effect of clear-cutting on streams and wildlife.

By now workers were descending on downtown Los Angeles. People in agriculturally related industries were tackling a multitude of office chores, ranging from sorting and delivering mail, programming computers, and interviewing job applicants, to purchasing supplies, preparing legal briefs, and selling truckloads of fruits and vegetables. At an agricultural college, professors were busy teaching while researchers prepared for a new round of experiments.

What, you will ask, is the significance of these diverse activities? Simply that they—and thousands of other jobs—are all part of the far-flung, vibrant, and challenging farm, ranch, and agricultural industry, a vast business which offers more diverse careers than almost any other! No matter what your interests are, there probably is an opening awaiting you somewhere in this field.

AGRICULTURE—A MOVING TARGET

One only need pick up the daily newspaper to see what a fluid field agriculture has become. Natural disasters like floods, excessive heat or cold, insect invasions, tornados, or hail can suddenly cut off part or all of a nation's food supply. This may spell starvation for the affected people and sudden prosperity for farmers and ranchers elsewhere who help make up the food deficit.

Destruction of the Irish staple potato crop, which brought famine and disease, created one of history's worst natural disasters from 1845 to 1849. Almost 150 years later, *Phytophthora infestans,* the same tiny insect responsible for the earlier tragedy, was devastating potato fields in North America. Not only could this ruin farmers, but the transportation industry—trucks, railroads, ships—would also lose money. In American and Canadian commodities markets, traders would be desperate to fill potato orders, while supermarkets and other stores would lose revenue. A cry for help went out to entomologists and others concerned with protecting our food supply, and they worked night and day to develop an effective pesticide.

If you raised lambs for wool or grew cotton, news about Tencel would worry you. Courtaulds, a British chemical company, has produced a new fiber made from wood pulp. The fiber looks and feels like silk, is as strong as polyester and as absorbent as cotton or wool. On the other hand, while it threatens "King Cotton" and wool, it should benefit those in the lumber industry.

A news item in the *New York Times* in January 1995 reported that the wealthy Brazilian entrepreneur Olacyr de Moraes had built a private railroad across his country. He plans to ship soybeans from his Amazon farms and ranches, which are as large as Rhode Island, to the nearest Brazilian port, and will easily compete with American and Canadian farmers. Undoubtedly this development will upset world soybean markets and the business of countless soybean growers.

An exciting agricultural development which is certain to expand further is the increasing use of plants as raw materials in many manufactured goods. They will replace scarce and less acceptable petrochemicals. Some household items produced from plastics include adhesives, dyes, detergents, inks, pigments, plastics, cleansers, and wall paints.

Botanists have created high protein plants that grow profusely and may help create a "green food revolution" for underdeveloped nations subject to periodic famines. While some success has been achieved, there remains a real problem. Local eating habits mean that a rice-eating country wants only what its inhabitants are accustomed to consuming, rice, for example, and not wheat, corn, or the new food. Thus the answer is to increase production of native foods, not introduce others. Meanwhile millions are frequently on the verge of starvation.

After Kobe, Japan, suffered a tragic earthquake on January 18, 1995, and its port was temporarily closed, the world learned the importance of this facility to international trade, especially agriculture. During the previous year, records revealed that the five principal imports to Japan from the United States, expressed in number of containers (20' × 8' × 8'), were: animal feed, 35,000; chemicals, 31,900; paper products, 27,500; fabrics, 27,100; and meats, 20,200. All of these imports except chemicals were products of farms, ranches, and forests.

Reading the above, you can begin to see the career potentials in this intricate field. Agriculture not only provides our food, but is tied into our social and economic welfare as well.

THE DEVELOPING NATIONS

Although food shortages occur in many parts of the world from time to time, in recent years there has been a shift in production among developing countries. Some Asian nations produce enough food to export their surplus, whereas sub-Saharan Africa suffers the most serious food shortages. Many of these poverty-stricken countries are wracked by tribal and other internal wars. This threatens stability as well as the prospects for bringing in aid, be it economic, technical, or educational assistance. Careers in the developing world call for a sound education in a technical field, such as animal health, plant genetics, economics, and others, as well as skills in languages, human relations, and understanding other cultures. Future as well as present careers in agriculture will center on or evolve from agriculture's basic role in years ahead—production of food and fiber.

If an agricultural career abroad is your goal, we suggest that after you have completed your education you consider doing a stint in the Peace Corps. Your assignment may not be in your chosen field, but the service can provide you with valuable experience and enable you to decide whether working overseas would be a satisfactory career for you.

COMPENSATION

Earnings figures are difficult if not impossible to obtain in any industry. For competitive reasons, companies usually refuse to divulge them. Therefore we have to rely mostly on such limited data as is compiled by the Bureau of Labor Statistics.

However, we do know, for example, that some farm and ranch owners make over $36,000 annually; a forest products technologist or entomologist may earn upwards of $60,000 a year; and a good administrator or head of a large company that processes farm, ranch, or forest products, may realize $100,000 or more. Appendix B gives some income figures, educational requirements, and employment opportunities for selected agriculturally related occupations. Your high school or college guidance counselor may give you some better idea of what lies ahead for a particular job category.

It is interesting to note that in Carol Keiman's book, *The 100 Best Jobs for the 1990s and Beyond* (Dearborn Financial Publishing, Inc., 1992), the author lists the following agriculturally related careers: Accountant/Auditor, Agricultural Scientist, Biological Scientist, Chemist, Clerical Supervisor/Office Manager, Computer Systems Analyst, Environmental Scientist, Farm Manager, Food Scientist, Public Relations Scientist, and Wholesale Sales Representative.

It is safe to say that if you apply your talent and education to the agricultural field you will probably do as well, if not better, than in many other endeavors. Remember, too, that the most important thing is to be happy in your work and have a sense of accomplishment as well as a conviction that you are making an important contribution to the industry with which you are associated. Few fields offer such opportunities as farming and agriculture.

CHAPTER 2

CAREERS IN THE SUPPORT SECTOR

Did you know that to have a career in farming, ranching, or agriculture you don't need to ride a tractor, feed or water livestock, study the effects of pesticides, improve the packaging of some animal product, or hold one of the thousands of scientific positions discussed in the following chapters? The answer lies in this chapter. We start by asking what do a farm bureau, corporate ranch, food processor, stockyard, agricultural college, or nitrogen laboratory have in common? Support personnel: men and women who work mostly behind the scenes but without whom no group, company, agency, or organization could function.

They are the managers, computer operators, accountants, secretaries, public relations specialists, typists, file clerks, lawyers, personnel specialists, and laboratory technical staff, as well as the maintenance crews, vehicle drivers, mechanics, custodians, food service workers, and all the others who provide these and other vital ancillary services. Each of these people is essential to the operation of each and every organization, including all operations related to farming and agriculture. This should be good news to anyone who has no interest in milking cows, plowing fields, rounding up cattle, peering through a microscope, tramping through fields and forests, or studying the grasshopper's life cycle, but who would still like to be part of this, the world's largest industry.

As you read ahead you will find that almost all the positions discussed call for highly trained men and women, and often require a graduate degree. As you discover the innumerable fascinating activities associated with agriculture, perhaps you can visualize yourself becoming part of the team by qualifying and joining up as a member of the support group.

EMPLOYMENT OPPORTUNITIES

Here are some of the more common job classifications and job titles that comprise the usual support functions. No company, farm, or ranch operation will need all of these workers, but somewhere you should find an opening for your particular interest or skill.

Accounting. Accountant, bookkeeper, cashier, controller, credit manager, economic analyst, inventory controller, manager of accounts payable and receivable, payroll manger.

Administrative. Administrator, administrative assistant, assistant administrator, office manager.

Commodities Trader. Analyst, floor trader, manager.

Computer Services. Computer service director, computer specialist, computer specialist operator, computer services manager, computer technician, database coordinator, keypunch operator, programmer, word processor.

Engineering and Maintenance. Assistant engineer, carpenter, custodian, electrician, elevator repairman, engineer, groundskeeper, mason/plasterer, painter, plumber.

General Office Services. Clerk, executive secretary, mail clerk, receptionist, secretary, stenographer, switchboard operator, typist.

Library. Cataloger/classifier, librarian, reference librarian.

Marketing. Advertising specialist, customer service representative, market analyst, sales manager, salesperson.

Personnel. Employment interviewer, job analyst, personnel director, wage/salary analyst.

Public Relations. Photographer, public relations director, public relations specialist, writer.

Purchasing. Buyer, purchasing director, stock clerk, stockroom manager.

Technical Staff. Laboratory technician, technologist.

You can prepare for many of these positions at a two-year community or junior college or at a vocational technical school.

For information about these positions consult the suggested readings in appendix C. You will also find many of these jobs included in appendix B, "Statistics for Selected Positions in Agriculture."

CHAPTER 3

CAREERS IN FARMING AND PRODUCTION AGRICULTURE

If you have ever flown across the country coast-to-coast, no doubt the variety of farms and farmland that passed beneath you made a lasting impression. In New England and New York, dairy and vegetable farms dot the landscape; further south you would see cotton, tobacco, and vegetable farms, as well as orchards. The rich black Illinois soil would probably surprise you if you have never seen such farmland, and then the midwestern states show off their far-flung acres of corn, wheat, soy beans, alfalfa, and other crops. These grow on farms neatly laid out in apparently unending squares as far as the eye can see. Next, the Great Plains reveal ranches and grazing land, and then you cross over the majestic mountains which give way to the lush farms of California, Oregon, and Washington, with their variety of vegetables, fruits, and other crops.

You have been viewing the "production factories" of our most important industry, for farms not only feed every American and countless people overseas, but also produce products used in many industries.

You may have heard it said that the "farmers are on their way out," or "nobody wants to farm anymore," or that "you can't make money on a farm." Don't believe it! Farming and ranching are here to stay as long as there are people who must eat. Farming is adjusting to the times and many farmers make excellent incomes. Best of all, because of the opportunities it extends to those interested in this field, it attracts intelligent and ambitious young men and women and offers untold prospects to those who choose to embrace this way of life and prepare for a career in the agricultural sector.

AN IMPARTIAL LOOK AT FARMING

Over the past forty years, some 30,000 farms have closed annually. While there were 5,399,437 working farms in 1950, by 1992 only 1,925,300 has survived, the lowest number since 1960. At the same time, the average size of the farms still in business had increased and were producing more than ever before.

Large farms with annual sales of over $250,000 dominated the business, while operators with income ranging from $30,000 to $250,000 were experiencing more and more difficulty making a profit, let alone breaking even. Many of the these were operated by families who had other sources of income. The plight of the eastern dairy and other farmers was discussed at a Vermont Town Meeting which considered "The Human Toll on Agriculture" on May 21, 1994.

The most successful participant said he owned 900 cows and employed 12 hired hands to do the work. Although he was making money, his real concern was that he did not have enough time to spend with his family. Another farmer reported that he and his wife tended 90 milking cows. Every day they are up before five and finish their morning chores by eight. Once they have eaten breakfast, they return to the barn and work there until noon. No later than three they are back in the barn again and finish their evening chores "by seven if we're lucky. It's a sixteen-hour day. It used to be fun, but it's no fun anymore. It takes a toll on our backs, and it's a struggle just to pay the bills." A common complaint at the meeting was that the prices farmers are receiving for their milk is even less than ten years earlier. They are locked into a market that doesn't take account of inflation or other factors. "Take it or leave it," the buyers of milk and other farm products seem to be saying. Little wonder so many small farmers are giving up!

The picture is not all bad, however. Quite a few farmers who had 200 cows or more are making a good living. Some are growing specialty crops like berries, apples, or table corn. Others affirmed that their wives are working nearby in business or industry to earn the necessary added income which enables them to keep going. In some of the most successful farms, the women are involved, usually keeping the books, ordering supplies, and handling correspondence, but as farms grow larger, the women are less involved, probably because it is possible to hire farm workers.

Out on their farm in Belle Plaine, Minnesota, according to a picture caption in *Modern Maturity,* Bill and Florence Halquist, who have been married 38 years, tumble out of bed every morning, including Christmas, at quarter to three. They milk their 330 Holstein cows and again at three in the afternoon start their four-hour evening milking stint. Although they have not had a week off in 18 years, they declare that if they did not enjoy their routine they would not be doing it.

Dairy farming, whether in Vermont, Minnesota, or California, is a super-demanding business and makes other types of farming like Howard Sugarman's operation seem tame by comparison.

Another View of Farming

On October 15, 1994, Howard Sugarman drove his huge combine to his Illinois cornfield and steered it down the first eight rows, the rumbling machine ingesting the tall stalks and expertly stripping the kernels from each cob. The monster inched its way at a speed of four miles an hour. Every ten minutes one of the two farm workers drove alongside to offload the accumulated corn and whisk it to the barn, making it unnecessary to stop the combine. From time to time Sugarman checked the moisture analyzer to make certain that the corn was not too wet. That evening, by the time the last stalk had disappeared into the combine, Sugarman had cut and stored a record crop, but alas, he reflected, other farmers also had bumper crops and that would send prices plunging.

"When your farm is producing more than ever, you can end up with a loss because of over-production elsewhere," he lamented. "I have another field some 35 miles away. Fortunately that's in soybeans and I'll make a little money there." Sugarman had a third field of corn still further away, and this he handpicks for a corn processing company which pays a premium price. His goal for 1994 was to achieve sales of almost $800,000. When he closed his books at year-end he learned that his profit was over $200,000!

FARMING IS A CHALLENGE LIKE ANY OTHER BUSINESS

Granted Mr. Sugarman did not have to rise at 2:45 A.M. like the Halquists. Nevertheless every farmer has a steady routine of unending chores. On most farms each season makes its own demands, which may include one or more of the following:

- Spring calls for plowing, fertilizing, seeding, repairs.
- Summer requires hoeing, weeding, haying, some harvesting (depending on crops), to say nothing of painting and repairing buildings, and working on new projects.
- Autumn brings final harvesting, cleaning up fields, winterizing buildings and machinery.

- Winter is a time for repairing equipment, fixing fences, getting ready for spring, and planning the next year's operation.

Each day makes its demands too! In between the essential farm or livestock chores are those of bookkeeping, supervising workers, ordering and inventorying supplies, meeting with agricultural specialists and manufacturers' sales representatives, and making major and minor repairs.

If there is time, the farmer might squeeze in a trip to the county fair or some other short trip with the family.

"A farmer never lacks for something that has to be done," an elderly man with a smiling but weatherbeaten face, told us. "That's why we seldom leave the farm."

The same is true for farmers who raise livestock, including poultry. Those who specialize in selling eggs and chickens must endure a twenty-four-hour responsibility overseeing temperature control, feeding, watering, and gathering fresh eggs which must be graded, packaged, and sold.

Farmers who raise animals for dairy or meat products work constantly year-round. Animals must be fed and watered daily and cows must be milked twice a day. As we have seen, unless these operators can share their work with others, it is impossible to get away. Some types of farms, such as crop farms, require seasonal work. Here, owners and managers labor from sunup to sundown during planting and harvesting seasons, but they may work only six or seven months a year and many are able to hold down second jobs.

If it appears we are doing our best to discourage you, far from it! We are merely trying to give you an honest picture of what it means to enter this vocation. Farming is no different from many businesses which require long hours, hard work, a large investment, and years of previous training. Many doctors, plumbers, hardware store owners, and innkeepers work equally long hours and have comparable responsibilities and problems. What it all amounts to is that farming provides a different kind of challenge and way of life, and offers its own rewards to those whom it attracts.

FARMING IN TRANSITION

Mr. Sugarman's Illinois farm is typical of most large commercial operations, except the very extensive businesses owned by big corporations. The corporate farms have the advantage of being able to purchase expensive equipment and fertilizer at favorable prices, select the acreage they want to use each year to best advantage, and employ sufficient help. On the other hand, the small farmer has limited acreage, may have to use it year after year, often has insufficient help, and may lack the latest labor-saving equipment which can be very cost-effective. For these and other reasons small farms close, either because they cannot make enough money to pay their bills or because the owners are growing old and have become tired of the endless work. Such farmers are only too glad to sell out to land developers or others seeking their acreage.

You probably noted that Mr. Sugarman works three widely separated farmlands to assure that he has enough production to make his business profitable. No doubt part of the reason for his having to work three plots is the fact that today our farmland is vanishing at an alarming rate, with one-third of it declared endangered. That is because those 320 million acres are close to cities and are threatened by the continuous development that is occurring as a growing population requires more and more houses, the inevitable shopping malls, and the other amenities of modern life. It is not a happy prospect. As prime farmland disappears, it is urgent that the remaining acreage be used ever more efficiently and production be increased accordingly.

FARM MANAGERS AND OPERATORS

In 1992 there were approximately 1,218,000 farm managers and operators, the majority managing crop production and most of the balance managing livestock production. A few operated companies which specialized in contract harvesting and farm labor contracting. The following information will give you an idea of what these positions entail.

FARM MANAGERS

A farm manager, as the term *manager* implies, is in charge of farm operations, but he or she will not necessarily fit into a neat management slot. On a crop farm the manager may be responsible for the overall direction of the operation or may supervise one part of the business. On a livestock farm the manager may oversee the entire business or just a single activity, such as feeding the animals. A highly trained professional farm manager may work for a company that specializes in farm management and be assigned to manage one or more farms. His or her duties may include not only supervising the workers, but also establishing planned maintenance schedules for all buildings and equipment, monitoring production and marketing, establishing production goals, determining financial policy, and preparing periodic financial reports for the owners.

FARM OPERATORS

Farm operators are either owners or tenant farmers who rent the use of the land. They must direct the business and supervise the workers. In the case of most farms, the operator has one or two family members or hired employees to assist with the work. On some large farms, however, there may be as many as a hundred workers, including a bookkeeper, computer specialist, truck driver, and sales representative. The farm operator may spend much of his or her time meeting with the managers and supervisors in charge of various operations.

Qualifications

Farm managers and operators should have good management skills, including a knowledge of bookkeeping, accounting, computer operation, and sources of credit. Equally important is a familiarity with the latest government agricultural support programs and the paperwork required for this and other necessary business reports. Keeping up with agricultural technical reports and developments well as growing conditions and environmental matters is a must. Mechanical aptitude in using tools and fixing equipment is needed, especially on the small farm. Finally, a love of farming is an obvious requirement.

Training

Sound advice on training for a career in farming can be found in the *Occupational Outlook Handbook:*

> Growing up on a family farm and participating in agricultural programs for young people sponsored by the National FFA Organization or the 4-H youth educational programs are important sources of training for those interested in pursuing agriculture as a career. However, modern farming requires increasingly complex scientific, business, and financial decisions. Thus, even young people who have lived on farms must acquire a strong educational background. High school training should include courses in mathematics and the sciences. Completion of a 2-year and preferably a 4-year program in a college of agriculture is becoming increasingly important.
>
> Not all people who want to become a farm manager grew up on a farm. For these people, a bachelor's degree in agri-

culture is essential. In order to qualify for a farm manager position, they will need several years' work experience in many different phases of farm operation.

Students should select the college most appropriate to their specific interests and location. All States have land-grant universities that include a college of agriculture; their major programs of study include areas such as dairy science, agricultural economics and business, horticulture, crop and fruit science, soil science, and animal science. Also, colleges usually offer special programs of study covering products important to the area in which they are located, such as animal science programs at colleges in the Western and Plains States. Whatever one's interest, the college curriculum should include courses in farm production and in business, finance, and economics.

Farm operators and managers must have the managerial skills necessary to organize and operate a business. A basic knowledge of accounting and bookkeeping can be helpful in keeping financial records, and a knowledge of credit sources is essential. They also must keep abreast of complex safety regulations, requirements of government agricultural support programs, and paperwork faced by other small businesses. Familiarity with computers is important especially on large farms, where computers are often used for record-keeping and business analysis. For example, some farmers use personal computers connected to telephones to get the latest information on prices of farm products and other agricultural news.

Earnings

According to the U.S. Department of Agriculture, operators of cotton, fruit, vegetable, and poultry and egg farms earned an average cash income net of expenses of more than $100,000 in 1993. However, cattle and tobacco farms made less than $15,000 in cash income on the average.

In 1992 farm managers who were employed full time had median earnings of $382 a week. The middle half made between $382 and $545 a week. The lowest-paid made less that $185 weekly, with the highest-paid managers making over $696 a week.

THE PROS AND CONS OF FARMING

Advantages

- You can live and work away from urban and suburban congestion with its smog, crime, and other problems.
- This is a healthy outdoor life for those who do not hold office positions.
- If you own a farm, you are your own boss.
- This is a depression-proof industry, although the economics and your consequent benefits can be rough at times.
- Government subsidies and loans assist many farmers and help in acquiring independence.
- There can be opportunity for personal growth and financial enhancement.
- It can be an ideal environment in which to bring up a family.

Disadvantages

- There is stress in many farm jobs, as well as for managers and those who own their own businesses.
- The hours can be long, with little time off.
- The income a farmer realizes can be beyond his control.
- There may be inadequate income for the hours worked and the capital invested.
- A large amount of invested capital is needed to compete with the huge corporate farms.

- Farming is becoming big business, forcing many small farmers to withdraw.
- Farming is subject to unexpected ravages by insects, disease, weather, floods, droughts, fire, and other catastrophes.
- Large animals, as well as planting and harvesting machinery, can make work hazardous.
- On a small farm family members may have to find outside jobs to supplement the farm income.

WHAT ABOUT YOUR PLANS?

Say you are interested in owning your own farm. Believe it or not, there are many other men and women of all ages who look forward to such a life. According to a recent Gallup poll, 22 percent of those surveyed would prefer to live on a farm. If you are fortunate enough to live on your parent's farm and are in line to join the family business, somehow acquire or inherit it, you have no problem. However, if you have to purchase your own operation yourself, how can you do this?

The obvious answer is to work on farms during summer vacations to discover whether or not you are fitted for such a career and will enjoy it. Once you know this is for you, you must study at a vocational/technical school or community college, or—preferably—earn a bachelor's degree at an agricultural college. Then you will want to find a position with a large or corporate farm or ranch.

Now you start obtaining your practical experience and open a savings account which you hope (perhaps with your spouse's help) will grow sufficiently to provide a down payment on your own business. Obtaining a loan for the balance of your capital may not be easy, but it is possible to do so through one of the federal agencies which guarantee farm loans and mortgages. You may even find a bargain, such as a farm or ranch the owners are ready to sell at an affordable price. In addition, there may be a farm or ranch operator who will rent the business to you or take you on as an associate, eventually enabling you to purchase the property as you gradually invest more money in it. The local farm bureau or county agent is available to advise and help you, too.

MANPOWER CHANGES

Consider what is happening in Wisconsin, as reported by the *Wall Street Journal.* Between 1987 and 1994, the number of farms dropped by 9.5 percent. Recognizing the seriousness of the situation, the state's Agricultural Technology and Family Farm Institute installed a computer database to help those older farmers who want to retire find interested young men and women to succeed them.

In nearby Iowa, projections showed how the farm picture is changing. It was expected that only a third of the farmers' children will stay on the farms. The state was therefore trying to match young couples who aspired to have their own farms with those farmers anxious to sell out. Randy McAllister, 32, purchased a 128-acre farm with the help of the Iowa Farm-On program. Low-interest state loans assist such newcomers realize their dreams.

If you aspire to own a farm, contact the state agricultural department and ask what help is available to you.

WHAT ABOUT THE FUTURE?

It is always dangerous to predict the future, but based on current trends it seems safe to list the following anticipated changes:

1. The market for organically grown foods will continue to grow as increasing numbers of health-conscious Americans want these products.
2. Genetic engineering will become an important factor in farm production.
3. Livestock that has not received hormones will be in greater demand.
4. Farmers will become more market-oriented and respond to what consumer demand rather than continuing to produce the same crops or livestock they have always raised.
5. Technology will continue to change the ways farmers and ranchers run their operations, thanks to improved and more efficient machinery, methods of planting and harvesting, and livestock and office management. The computer will bring further changes and improvements in the form of labor-saving devices.

As already mentioned, mid-sized farms will disappear for the most part, as owners find it impossible to cope with the financial and other realities of farm operation. And overseas markets for grain and other foodstuffs may remain strong for a few years. Nevertheless, this demand is not assured, because other countries will improve their agricultural production.

We have introduced you to farms and ranches in order to expose you to the overall business and what you may expect to find if you enter it. With this background information you can better evaluate the diverse paths one can take in this field. In other words, we just outlined the business and its possibilities; now we shall put "meat on the bones" as we discuss the numerous careers that exist in production agriculture—the production of food and fibers, whether it be by growing crops, running a dairy farm, raising beef cattle and flocks of poultry, producing nearly pure cellulose clothing fiber, managing catfish production, or working in the verdant forests.

These production careers concentrate on using and managing human, capital, and natural resources, and often are so specialized that it becomes impractical to switch quickly or economically to another enterprise. For example, it would be difficult for the owner of an orange grove to change to another crop, or for a large grain grower to develop another business. Production agriculture, therefore, has two purposes: producing a product, and realizing a profit.

CROP PRODUCTION

At the core of these complex, specialized systems is production of basic carbohydrates and proteins in grains and legumes. Grains, generally, are classed as food grains (primarily wheat and rice) and feed grains (primarily corn and sorghum). When combined with soybeans, the largest planted crop for protein and oil, the acreage of these crops accounts for two-thirds to three-fourths of the harvested cropland in the United States.

This abundant supply of crops makes possible the complex, high-volume systems of meat production, whether in broiler houses in Georgia or in dairy farms of California or beef feedlots in Texas. From the effort devoted to producing these staple crops comes most of the carbohydrate and basic source of protein for our daily breakfast cereals or meals at fast-food establishments. In a sense, crop production is the most basic activity in agriculture, utilizing natural resources of soil, sunlight, air, and water to produce carbohydrates and proteins, the two key components in the food chain.

Understandably, the range of crop production careers is wide:

- Agricultural engineer
- Agronomist
- Cash grain farmer
- Citrus grower
- Cotton producer
- Crop consultant

- Entomologist
- Farmer
- Fruit grower
- Irrigation specialist
- Nut orchardist
- Peanut producer
- Plant pathologist
- Rice farmer
- Seed producer
- Service representative
- Tobacco grower
- Tree, vine fruit grower
- Turf grass sod producer
- Vegetable grower

Few careers in production agriculture fit a single, neat package. A crop consultant may have an education in plant pathology, but may serve primarily as an irrigation specialist. Or a person who majored in agricultural communications may be a service representative to customers of a herbicide manufacturer. Whatever the career in production agriculture, it involves producing plants from soil (and in rare instances from nutrient solution by hydroponics) and poultry, livestock, or fish from the harvest of plants produced.

As careers in production agriculture are largely located in the world of biology, it is logical for the education for most of these careers to be in areas of biological sciences. Sometimes an education in business, financial management, or the like is augmented with biological sciences. However attained, successful careers in agriculture production need the skills of technology and business management in some combination.

Future careers in production agriculture will place an even greater emphasis on profitability and competitiveness in international markets. Low-priced commodities, such as grains, dictate low per-unit production costs to provide a margin of profit.

CROP PROTECTION

Teaming up with agronomists is a career group of specialists trained to protect plants from a host of pests—hence, the term pesticide or crop protection chemicals. This line of products contains both biological and commercial chemical products to control nematodes, bacterial and fungal diseases, a broad range of insects, and hundreds of weeds that cut crop yields and crop producer profits.

Methods of plant protection are shifting to biological and nontoxic methods. Examples include biological parasites and sex-lure baits as well as viruses, attractants, sterility genes, and predators. Given the current public concern on the impact of commercial pesticides on the environment, careers in plant protection will no doubt focus more on these methods than in the past.

Plant protection specialists also play a key role in the production of horticultural and vegetable crops. Quality is especially important for these tree and vine fruits and vegetables. Consumers have become accustomed to insect- and disease-free fruits and vegetables, and such quality is possible only through the work of career specialists in plant protection methods.

Opportunities for careers in crop production range from pineapple in Hawaii to potatoes in Maine. The desert is no exception. With current irrigation methods, yields of citrus, cotton, vegetables, grain, and other commercial crops are some of the highest in the world.

In urban areas, horticultural careers in turf-grass production, lawn and ornamental care, and floriculture offer an expanding number of opportunities. While agronomic crops offer the largest use of germplasm (a term used collectively for genes in seeds and in plants themselves) transfer by seed, commercial plant nurseries reproduce more plants asexually through grafting and layering techniques than any other group. Without specialists in reproducing and growing such ornamental plants, there would be few plants available to enhance the appearance of the yards and patios of American homes or the landscaped areas of commercial areas, parks, and playgrounds.

There is no set mold to show the usual shape of interests and aptitudes of individuals for various careers in production agriculture. A few clustered preferences or interests may give a clue to such careers.

- Agricultural engineering (mechanization)—interest in designing buildings, operation of equipment, mechanics of water transport and irrigation, and materials-handling systems.
- Agronomy (soil and crop sciences)—natural curiosity in applied science, growth processes of plants, chemical and biological properties of soils and plants, nutrition and physiology of plants, plant genetics, and molecular biology.
- Entomology—interest in insect physiology, reproduction, chemistry of metabolism, ecological factors affecting insect prevalence, principles of toxicology, and insect genetics.
- Horticulture—many interests are common to those listed above for agronomists, combined with special interest in fruits, vegetables, and floriculture.
- Plant pathology—interest in biochemistry, botany, epidemiology, molecular biology, plant physiology and anatomy, and microbiology.

LIVESTOCK AND DAIRYING

Americans and Europeans obtain about 40 percent of their dietary calories from meat products. Beef and dairy animals, with their multiple-compartment stomachs, are some of nature's marvelous converters. They consume cellulose, lignin, and other carbonaceous substances in otherwise useless forage plants on the range, and produce meat and milk. They consume non-protein nitrogen, such as urea in feed supplements, and produce protein.

Thus, careers in beef production extend from involvement with calving operations on ranches and farms to pharmaceutical laboratories where growth hormones or medication products are developed. A common objective in these careers is to improve conversion of plant material to meat, that is, to protein.

Careers with cow/calf operations range from herd manager to geneticist. Such careers include specialists in developing and marketing controls for a host of pests—from intestinal parasites to face flies—and, obviously, those with basic veterinarian skills, whether by education or on-the-job experience.

Genetic research in the beef industry is in the high-tech arena. It includes extensive computer records and, as in dairying, embryo transplant technology. This high-tech system seeks out not only high growth-rate animals but also animals that produce a quality of beef to meet consumer preferences.

Feedlot operations, generally located in the Great Plains states, have some of the largest logistical requirements in all of production agriculture. Careers in such operations include purchasing agents of millions of dollars worth of cattle feed and ration (feed) specialists who seek to minimize costs of purchased grains to pest-control professionals who work at single sites containing up to 20,000 animals. New feed supplements and medical products require a depth of nutritional and medical knowledge. And, as in land management, feedlot operations use computer facilities very extensively.

An adjunct to many animal-science departments in the land-grant universities is the equestrian career. In this, students learn anatomy, physiology, nutrition, and management of horses, much as do those preparing for an animal-science career. Management of horse stables and riding schools is a rapidly growing career area in commercial recreation, and an animal-science major can be one of the educational backgrounds for this.

Careers in dairying, as in beef production, are highly science-oriented. These careers involve lactating animals that produce one of the most wholesome and widely used human foods. Milk is a primary source of calcium to meet the recommended dietary allowance (RDA) of 1 gram of calcium per day. Milk also is the principal source of vitamin D for humans, and with its sugar lactose and protein casein, it continues as a major part of our diet.

As a result of more intensive farm operations and daily contact, dairy operators give individual animals much more attention than is possible with beef. Career preparation in dairy science, therefore, centers on assessing individual animal performance, including calf growth rates, daily milk production, analysis of milk, and period of lactation. Those in such careers utilize the tools of modern nutrition, genetics, physiology, and parasite control to increase production per cow. The final goal is to realize the best possible profit margin in an industry under increased price and cost pressures.

Intensive animal care in modern dairies presents special opportunities for careers in animal welfare. In nutrition, there are special computer programs to calculate optimum feed ingredients combinations with minimized costs. There are rumen buffers to adjust the acidity in the rumen, the first stomach of the cow, and there are special mineral and medicated feed supplements to include for an individual animal.

With such a specialized operation at a modern dairy, it is understandable that a whole array of unique equipment also is necessary. In dairy science, some careers deal solely with machinery for milking and on-farm cooling and storage systems. Use of "misters" to cool dairy cattle in states such as California and Arizona is growing in popularity as a means of reducing milk-yield losses during periods of high heat. The misting systems, operated by automatic controls, lower the body temperature of dairy animals and significantly help the animals maintain high levels of milk production despite summer heat.

There is a large career field with pharmaceuticals associated with dairy science. Grubs, roundworms, and fly larvae need to be controlled in a modern dairy. This control is possible only with health management practices combined with chemical pest control.

Another fast-growth career field in dairy science is in cattle reproduction. Artificial insemination is the common breeding practice today, providing the advantage of access to high-producing genes from breed associations. In the biotechnology field, dairy science is leading with embryo transfer, or transplant.

Examples of career opportunities in beef and dairy science involved in production agriculture are:

- Animal breeder
- Cattle rancher
- Dairy farmer
- Dairy herdsman
- Farm manager
- Horse rancher
- Livestock producer

POULTRY AND AQUACULTURE

Few if any sectors of production agriculture can equal the growth record of the poultry industry since World War II. In 1950 the average American ate 25 pounds of poultry and the industry produced 3.1 billion liveweight pounds. By 1992 per pound consumption was over 100, with 28.7 billion liveweight pounds produced. Career opportunities are concentrated into three groups:

1. Science—research, teaching, extension
2. Business—management, marketing, finance
3. Technology—production, field work

This complex, integrated industry offers career opportunities throughout its five tiers:

1. Breeder farms to get the right genes in the eggs
2. Commercial egg-producing farms to provide the eggs
3. Hatcheries to provide the chicks
4. "Grow-out" farms to produce the birds

5. Processing plants to package the poultry products for consumers

A relatively new form of production agriculture is aquaculture. It involves producing fish, such as catfish, bass, and shrimp, in confined ponds utilizing algal biomass as food. Fertilizer for aquaculture can range from a commercially produced product to chicken manure and chicken feed.

This development offers opportunities in careers specializing in marine life. Operations for shrimp will be limited to the southern coastal states since they require water temperature of at least 68 °F.

Examples of careers in poultry and aquaculture are:

- Aquaculturalist
- Breeder farm operator
- Business manager
- Commercial pond operator
- Computer specialist
- Equipment design engineer
- Equipment manufacturer
- Feed research director
- Field representative (feed, pharmaceuticals, etc.)
- Fish nursery operator
- Hatchery operator
- Poultry geneticist
- Poultry health specialist
- Production manager
- Sales and marketing

FORESTRY

The Ohio State University has reported that printing one Sunday issue of the *New York Times* consumes pulp from more than 150 acres. Yet, such newsprint is only one of some 5,000 products created from forest resources.

Careers in forestry management, genetics, biology, and many other fields in forestry provide this large number of products, and in large volume. Additionally, there are careers specializing in forest ecology to help improve forest production of wildlife in wilderness areas and to monitor the effects of logging near dam-reservoir sites.

In forest nurseries, geneticists are splicing genes and developing "super trees" for tomorrow. The nursery specialists produce seedlings for reforestation, combining as many traits as possible for insect and disease resistance and for fast growth. Many farmers, ranchers, and urban managers depend upon such nurseries for their forest stock.

Specialists in forestry careers also include many in research. In addition to adding to our knowledge of hereditary factors accounting for cellulose and lignin characteristics in the tree, there are career foresters using enhanced satellite imagery to study forest cover in large geographic regions. Through this technology, they monitor forest growth and plan harvesting and reforestation operations.

Many careers in forestry are with private industry. Government agencies employing specialists in forestry include USDA's Forest Service, Soil Conservation Service, and the Tennessee Valley Authority. Examples of these careers are:

- Forest ecologist
- Forest geneticist
- Forest ranger
- Forester
- Logging operations manager
- Lumber mill operator
- Park manager
- Range conservationist
- Satellite imagery specialist
- Soil conservationist
- Wildlife specialist

CHAPTER 4

ENHANCING OUR FOODS AND FIBERS

The complexity and variety of careers in industries that process food and fiber far exceed those in production agriculture. These career opportunities extend from managers of plants that process more than 8,000 broilers per hour with as few as 36 employees to positions within the frozen food industry—an industry that freezes more than 12 million tons of food annually.

People with careers in food processing make possible a system that enables one to eat what one has not produced. These careers bridge the gap between production of basic food commodities and attractive packages of food, shrink-wrapped and ready for the consumer. They range from flour-milling engineers to chemists who can manipulate protein enzymes in doughnuts.

Food processing has several fundamental functions, including conversion of products to enhance nutritional value, preservation of quality, and protection against destructive agents such as molds and bacteria. With proper conversion, preservation, and protection, consumers have wholesome food available for many seasons following production which may have occurred in locations far distant from that of final consumption.

Central to the need for food processing is the fact that few foods in their original, natural state are free from substances that quickly deteriorate quality. Through a variety of highly technical processes—including refrigeration, fermentation, heating, dehydration, and chemical sanitation—these substances and agents are removed, destroyed, or arrested in development.

Such a feat calls for the expertise of food technologists, engineers, and chemists to utilize fully the knowledge gathered from a vast array of scientific fields, including chemistry, microbiology, physics, engineering, and toxicology. There are no fixed boundaries describing career opportunities in food and fiber processing and related agribusiness. Not surprisingly, the large majority of these careers occurs within the private sector. Clearly, food processing is an information-intensive industry.

CROP PRODUCTS

According to an account by the Chicago Board of Trade, wheat cultivation dates back 7,000 years (most likely to Eastern Europe or the Near East), and soybeans to at least 5,000 years ago in China. Since the earliest rudimentary processing, such crops have provided carbohydrates, protein, and minerals for human sustenance. Fundamentally, little has changed. Food processing technologists are still manipulating starches and proteins.

In terms of volume, grains lead all processed food products. The average annual U.S. consumption of wheat flour, alone, is 115 pounds per person—a product of only about 200 wheat mills. Grain processors include:

- Alcohol manufacturers
- Dry-corn processors
- Flour millers
- Malters
- Mixed-feed manufacturers
- Oat millers
- Soybean processors
- Wet-corn processors

INVOLVEMENT OF LIFE CHEMISTRY

The chemistry of life itself constitutes an integral part of many careers in the varied food-processing industries. These careers may involve the chemistry of acid hydrolysis, or conversion, of starch into sucrose (a simple sugar carried in human blood) or the chemistry involved in development of food-coloring substances.

The story of Kirchoff, a Russian chemist, is interesting. During the Napoleonic Wars there was a shortage of sugar in France and Germany as a result of the British blockade. Although there were special efforts at the time to develop sugar substitutes, in 1811 Kirchoff accidentally discovered that an overcooked mixture of potato starch and sulfuric acid produced a sweet syrupy substance. He had discovered that an acid hydrolyzes starch into sugars, much like the hydrolysis of starch into sugars by our stomach acids. A year later, Germany had three plants producing sweeteners from potato starch. For the discovery, the emperor of Russia awarded Kirchoff a lifetime pension.

Today, food processors provide careers involved in converting raw commodities into nutritious food products, such as the clear, colorless, liquid high-fructose corn syrup (HFCS). This product is shipped by trucks or railcars to soft-drink manufacturers, the predominant user of this "engineered" food product.

Corn sweeteners such as HFCS are less costly than sugar and now account for more than half of the caloric sweeteners used in the United States. This industry uses approximately 1.2 billion bushels of corn (more than 10 percent of domestic production) to produce 21 billion pounds of sweeteners, including 5.5 billion pounds of corn syrup.

Careers in starch processing include many others in addition to those dealing with starch hydrolysis. For example, there are careers in food fermentation. In this case, food processing careers include specialists in malting—the conversion of starch to sugar by enzymes in germinating grain—to specialists in fermentation—the conversion of sugars to alcohol by yeast.

Fermentation, as was the case with acid hydrolysis, was most likely an accidental discovery, probably by the Arabians. The word *alcohol* is derived from the Arabic word *alhuhi,* and alcoholic drink was produced from fermented barley centuries before Christ.

PROCESSING PROTEIN AND OILS

Processing of grains goes beyond conversion of starch grains of seeds. There are also many careers in processing protein and oils of seeds.

For examples of the careers in food processing that deal with proteins, one turns again to wheat—from which comes the only flour that contains the proteins gliadin and glutenin. Chemists and others in the baking industry understand that when wheat flour is moistened and kneaded, these proteins form gluten. This substance is elastic and entraps gases produced during fermentation of dough to enable bread to rise while cooking.

Good "bread wheat," therefore, must have a fairly high protein content. Low-protein wheat is used for pastry, crackers, and cookies. For products such as macaroni and noodles, food technologists use a low-protein white wheat called durum.

The major crop processed for protein is soybeans. A 60-pound bushel of soybeans yields approximately 40 pounds of a high protein (47 to 48 percent) feed for poultry and livestock.

The other important constituent of soybeans is oil. Today's soybean-processing industry obtains about 11 pounds of oil from each bushel of soybeans, in addition to the protein meal. The industry uses hexane as a solvent to extract the oil from the seed.

This oil solvent is also used with other seeds, including corn, cottonseed, and sunflower.

The processing of vegetable oils yields a number of career opportunities, for such oils are used widely in processed foods, including mayonnaise, salad dressing, margarine, and vegetable shortenings. Some oils also are used in a variety of industrial products, ranging from adhesives to paints.

Careers in processing crop products include:

- Bakery specialist
- Business management
- Cereal chemist
- Chemist
- Federal grain inspector
- Feed specialist
- Flour mill engineer
- Food engineer
- Food technologist
- Grain buyer
- Malter
- Microbiologist
- Nutritionist
- Oil mill supervisor
- Quality control supervisor
- Researcher

MEAT PRODUCTS

Careers in meat processing, as in crop processing, are knowledge intensive, with many being highly specialized.

In the United States, the average person gets about 40 percent of their daily calories from meat products. Thus, the volume of these products and the variety of processing methods are large. The number of career opportunities is large as well, from those involving enzyme chemistry to those concerned with packaging techniques.

The case of the broiler industry illustrates the numerous career opportunities in processing meat products. Managers of such processing plants are responsible for receiving truckloads of live birds or animals and converting these animals into wholesome products, attractively packaged for the consumer.

Under the manager's supervision are professionals and skilled workers operating mass production machinery. Such a combination of resources permits, for example, as few as 36 employees in a single broiler processing plant to process 8,400 birds per hour (140 per minute). Engineering of specialized equipment for such plants is a major industry in itself that offers engineering, promotion, and sales careers.

Many careers in food processing deal with protection of processed meats. Bacteriologists work in the meat industry because there is a host of pathogens that endanger meat quality. There are even "cold-loving" bacteria—psychrophiles—that must be controlled to protect meat in refrigeration. There are other bacteria that break down fats and proteins, resulting in undesirable odors and flavors.

Such microbes occur throughout nature—in the air in processing plants and in the wash water. Just as fresh vegetables begin to lose quality at harvest, so meat becomes a target for bacteria and other attack agents at slaughter.

Protecting of meat against such agents involves the efforts of specialists in heat treatment (boiling, cooking), freezing or chilling, physical barriers (packaging), and irradiation. Heating and/or freezing will destroy or arrest the activity of most destructive agents, but constant vigilance is necessary to prevent re-entry.

There are many careers in the meat-processing industry that demand a strong chemistry background. These specialists work with nutritionists, marketers, and administrative staff to develop products that have high quality, consumer convenience, and longer shelf life. Chemicals, such as sorbic acid and potassium sorbate, are used along with temperature (heating or refrigeration) to control molds, yeasts, and bacteria.

Other food chemists may specialize in flavorings and in the chemical make-up of the meat products themselves. They may deal with the chemistry of fats in chicken skins or animal fats as related to cholesterol in human diets. They may develop flavoring compounds or compounds to mask, or enhance, natural flavors, such as the use of monosodium glutamate in masking the flavor of mutton.

Specialists such as these work on age-old problems that cause some meats to be too tender, some to be too tough. Their work leads to an understanding of muscle chemistry and to the knowledge, for example, that a single compound, collagen, contributes to higher tensile strength of aging muscle tissue—and to tougher meat.

Meat inspection is another specialized career opportunity, whether for poultry, red meat, or fish. These inspectors, working for the USDA, inspect and grade a very large percentage of all commercially marketed meats. For example, 95 percent of all turkeys are inspected by the USDA. These carry an inspection imprint on the wrapping and a seal indicating the USDA grade.

The ranks of the meat-processing industry also include nutritionists. They work with food technologists on matters such as substituting potassium chloride for sodium chloride in boneless dry-cured ham as a means of reducing sodium in diets for those with hypertension (high blood pressure).

COMPETITION SPURS DEVELOPMENT

Competition in the food-processing industry for tasty, wholesome food and an increasing number of convenience products for the consumer has spurred a major effort within the industry for new product development, whether chicken "nuggets" or boneless, vacuum-packed ham slices. Product and packaging innovation in the food-processing industry is on the "fast track."

Convenience to the consumer continues to be a major driving force in new product development. Some of these developments point towards "freezer freedom"—preservation methods that do not require refrigeration. In recent years, boneless chicken products have overwhelmed packaged frozen chicken, an example of one product overtaking and replacing another in the marketplace.

Food scientists, teaming up with packaging specialists, have developed precooked foods that are vacuum-packed, sealed, and have a shelf life of 18 months without refrigeration. Such freezer-free foods save the expense of refrigeration for the trucker, the warehouseperson, the camper, the supermarket, and the consumer. And, package designers for this consumer-minded industry provide built-in cooking vessels. With the precooked product inside, these packages become meals after only a few minutes in a microwave oven.

Careers in processing fish products are expanding as well. Commercial farms for trout and catfish are providing mass production for mass processing. Specialists with careers in this processing industry help provide consumers with an increasing variety of fish products.

As is the case with poultry and the red-meat industry, an increasing percentage of these consumer products are deboned, ready-to-cook, and convenient.

With the need for increased technology, engineering careers are vital to the food-processing industry. Highly complicated equipment is required, whether for stunning birds prior to slaughter or for de-boning carcasses. Furthermore, there exists equipment specialization for each major meat product—poultry, pork, fish, and beef. There are a variety of mechanically de-boned meat products now available because of the work of those in these engineering careers. Indeed, "engineered foods" will provide many of tomorrow's new food products.

Packaging is another highly specialized career field in meat processing. Although appeal to the consumer is important, protection of the product is tantamount—even protection against tampering. Providing a barrier to airborne bacteria and molds is a basic requirement. Excluding oxygen is critical. Special plastics do this; sealed glass is perfect, but cost advantages and consumer preference dictate an expanding use of plastics.

Examples of careers in meat product processing are:

- Bacteriologist
- Business manager
- Chemist
- Engineer
- Food & Drug Administration inspector
- Food processing supervisor
- Food technologist
- Irradiation specialist
- Meat inspector (USDA)
- Microbiologist
- Packaging specialist
- Plant manager
- Nutritionist
- Quality control supervisor
- Researcher
- Technical sales representative
- Toxicologist

DAIRY FOODS

Milk is one of the largest volume products in the food system. In addition to milk itself, there are many dairy foods produced by the milk-processing industry.

The first step in this industry, which includes a wide variety of professional careers, is the collection of milk at the farm. Milk trucks, which may be large stainless-steel tankers holding 6,000 to 7,000 gallons of milk, represent the first handling step in the processing of milk.

At this step, the dairy food is a suspension of fats, oils, protein, sugar (lactose), and minerals. It also is considered "contaminated" at this time, because of the presence of bacteria from the milking process, as well as from on-farm storage. Neither of these processes can be absolutely aseptic, free from contamination of bacteria that are naturally present in air and water.

Thus begins the battle with milk-bearing bacteria that will continue until the product is consumed. Again, many of the careers in this industry center on protecting the dairy product—destroying or arresting germs that deteriorate quality. The principal protective means is temperature. Low temperatures arrest the activity of bacteria; high temperatures kill most bacteria of concern.

Pasteurization

The earliest career specialist in milk processing, and surely the most famous, was Louis Pasteur, a French chemist. While studying the chemistry of wine fermentation in 1864, he discovered that the heating of milk to about 145 ° F. greatly extended storage life. Only later was it understood that heat killed the bacteria that produce an enzyme, lipase, that converts the sugar in milk, lactose, to lactic acid. It also converts milk fat into its major components—glycerol (the basic fat substance) and free fatty acids.

Although the lines of division are not absolute, careers in dairy manufacturing, or milk processing, concentrate on work with milk and milk products.

In milk processing, sanitation specialists begin at the farm with requirements for milking operations and on-farm storage. Cleanliness of equipment and refrigeration are the first essential steps in arresting bacterial activity.

At the milk-processing plant, storage and bulk-handling specialists take charge. Bulk milk is stored in large silos refrigerated to 40 ° to 45 ° F. From these silos, holding about 25,000 gallons, milk begins its specialized-equipment treatment processes. The design and construction of this equipment offers engineers special opportunities.

With strong consumer preference for low-fat milk, separation of fat is usually the first treatment of milk at the processing plant. Through large electrically powered centrifuges, or separators, fat is separated from the fluid. Whole milk, typically, contains about 3.7 percent fat. By altering this, several classes of milk are produced.

Once the desired fat content is obtained, pasteurization begins. Career specialists in this operation have developed techniques for cutting time of heating by increasing the temperature. At 145 °F, milk must be heated for 30 minutes to kill bacteria. For the same pasteurization effect, milk may be heated for 15 seconds at 161 °F. The latter method is prevalent in large milk-processing plants today.

Such pasteurization methods, however, still require refrigeration. Another combination of heat and time, called ultra-high temperature (UHT), developed fairly recently, involves heating at 191 °F for one second, or 201 °F for 0.1 second. Combined with special aseptic package processes and materials, UHT milk has broken the "refrigerator claim."

This relatively new product can be stored on the shelf for six months without spoilage. Such high heat, however, alters the taste of milk, giving it a slightly "cooked" flavor resulting from the combination of protein and sugars. For certain uses, however, and especially in countries without sufficient refrigeration, UHT milk offers major advantages.

As in food-producing industries, milk products are under continuous development. The UHT product is one example. Another is a calcium-fortified milk tailored by dairy manufacturing specialists for special consumer needs, such as pregnant women and elderly persons with special calcium requirements. Such a product reflects the joint efforts of careers in production technology, human nutrition, and customer analysis.

Dairy manufacturing includes many products other than fluid milk, such as ice cream and cultured products. Each of these requires in-depth specialization for careers in their manufacture.

In production of cultured products (such as sour cream, yogurt, and cheese), there are unique career opportunities for those specializing in bacteriology, microbiology of yeasts, flavors, and other areas. Very specific strains of bacteria are selected for cultured products, depending on desired color, texture, and taste. Such cultures are available from specialized suppliers, or from in-plant "mother" cultures.

Those with career interests in chemistry also find opportunities in the dairy-manufacturing industry in production of cultured products. Extensive chemical monitoring is required of biological and chemical processes for these products.

There are also careers in this industry for microbiologists specializing in yeasts. For example, molds are necessary for certain cheeses, such as blue cheese. (Blue cheese is blue because of the specific mold.)

Representative careers in milk processing or dairy manufacturing are:

- Bacteriologist
- Cheese plant supervisor
- Dairy foods chemist
- Engineer
- Laboratory analyst
- Microbiologist
- Milk plant supervisor
- Packaging specialist
- Quality control supervisor
- Sanitation supervisor

FIBER

The industry necessary to convert cotton fiber into cotton fabric is one of the largest U.S. industries that process a single agricultural product. A much smaller industry exists for wool, but the case for cotton stands out in offering career opportunities.

Cotton is an unusual fiber. It is nearly pure cellulose, consisting of a linear, or chain-like, arrangement of certain modified molecules of glucose. Chemists call cellulose a linear polysaccharide.

Textile scientists have dissected the fiber so that they understand not only its chemistry but its structure. They know that it is a hollow fiber and that each fiber is simply an extended cell of the cottonseed surface in which nature deposits cellulose by an unknown process.

Processing cotton involves another product—the seed. It is a seed high in protein and oil. After oil is extracted, in a process similar to that for soybeans, there is the remaining meal. The oil goes into a number of food products, such as mayonnaise and margarine; the meal is used, primarily, as a protein source of poultry and livestock feeds.

With this range of products, there exist numerous careers in industries based on this one agricultural crop. A typical profile of products from one crop of cotton is shown by a report from the National Cotton Council.

- Eleven million bales (480 pounds each) are used by textile mills:
  - 64 percent goes to apparel
  - 30 percent goes into home furnishings
  - 6 percent goes into industrial products
- Four million tons of cottonseed:
  - 1.8 million tons of cottonseed meal
  - 169 million gallons of cottonseed oil

Operators of cotton gins are the first processors and often the first buyers. Or the farmer may sell the product to a local merchant or buyer who in turn sells it to a textile mill. The ginner usually buys the cottonseed and sells it to an oil mill. Typically, there are 162 pounds of seed for each 100 pounds of lint.

Grading of the lint follows ginning and largely determines its ultimate use. Experts grade each bale and measure its "staple," or length of fiber. This is just one career opportunity in the cotton industry.

Since cotton is a nonperishable crop, baled cotton can be stored for months in warehouses, either by farmers or other owners.

Persons in specialized careers oversee the processing of both the lint and seed of cotton. The first process for lint involves yarn production, a highly mechanized operation. Even the opening of bales has been fully automated.

Textile specialists blend cotton from different bales, based on staple and grades. They use computers to group the bales according to size, grade, and other characteristics.

The next step after blending the lint from selected bales is to feed the blended lint to carding machines, which produce yarn. There, textile specialists supervise the combing, straightening, and shaping of fibers into a thin web, which is molded into soft, rope-like strands called "slivers." With further processing on a spinning frame, yarn is wound onto bobbins.

Forming fabrics is the next step for the textile mills. At this phase, different textile specialists, chemists and engineers, produce a variety of fabrics, from the initial "grey goods" to the bleached, preshrunk, dyed, and printed fabric.

Computer specialists assist with an increasing number of operations in cotton-processing operations, from inventory control of individual bales to programming designs for color printing.

"Finishing" is the final step in fabric production. Cotton is one of the most versatile fibers and can receive hundreds of mechanical and chemical finishes. Textile specialists can change the appearance and feel of the fabric with these finishes or add special characteristics, such as durable press, water repellency, or shrinkage control.

A profile of the U.S. cotton industry, in addition to the 38,000 farmers who grow cotton, in 1994 by the National Cotton Council is shown below.

	Number of U.S. Businesses	Number of Jobs
Gins	1584	33,065
Merchants	286	2,855
Warehouses	751	7,729
Cottonseed oil mills	40	2,550
Textile Mills	802	160,526
Total	3463	206,725

CHAPTER 5

INDUSTRIES THAT HELP THE FARMER

The record of production agriculture—where one farmer produces food for 80 to 90 persons—is possible only because a number of other industries provide critical input. Our modern system of food production relies heavily on these input industries, for the farmer produces no diesel tractors for tilling the land, no chemicals to kill destructive insects, no medicated feed additives to control animal diseases, and no ammonium phosphate to provide nitrogen and phosphorus for plant growth.

Instead, the agricultural engineer, the chemist developing an ovicide (chemical to kill the eggs of insects), the animal physiologist developing synthetic proteins to improve cattle growth rates, or the chemical engineer producing fertilizers are examples of those in industry careers that provide critical farming inputs. These industries, in effect, provide labor and land substitutes in production of food and fiber.

Without mechanization, more laborers would be required to produce crops. Without chemical weed control, there would be more mechanical weed control needed, including hoes. Without nutrients in commercial fertilizers, acreage of land needed would increase sharply as yields per acre would drop to a fraction of current levels.

Such is the role of a host of agricultural production input industries. Those that provide equipment, crop inputs, and animal products illustrate the types of career opportunities in businesses that supply agriculture with crop and livestock production tools. An illustration of the magnitude of annual farm expenditures for inputs from such industries for just a single crop, cotton, was recently reported by the National Cotton Council:

- $500 million—agricultural chemicals (pesticides, defoliants)
- $450 million—fuel, equipment
- $300 million—fertilizers
- $100 million—seed

- $1,350 million—total for purchased inputs, excluding purchased labor

MECHANIZATION

The powered equipment on today's farms, ranging from tractors and planters, which place seed in the furrow, to electrical motors, which run self-feeding systems in broiler houses, is available because of those with careers in engineering. Careers for engineers also extend to the extensive industries that store and process food and fiber products.

With backgrounds in mathematics and physics, many engineers are involved in designing and testing farm equipment. Extensive work is necessary with prototypes and experimental units before a planter or combine can be delivered to the yard of a local farm equipment supplier.

Careers in engineering require first-hand awareness of new developments in agriculture. Farm equipment often has complex designs to enable multiple operations such as the simultaneous application of pesticides and fertilizers during seed planting. Each crop has unique requirements, whether for seed size and spacing, tillage, or harvest.

From the farm tractor manufacturer, the farmer today can select units that range in power from 50 to 370 horsepower. These pieces of equipment have specialty designs such as four-wheel drive, high clearance, and low and narrow profiles adapted to individual crops.

Numerous modifications in farm equipment extend from powered cherry pickers and tomato harvesters to tractor-type pluckers that pick up and crate live broilers. Again, a primary purpose of such specialized equipment is to use machinery in lieu of human labor. Careers in this engineering field are highly specialized, depending on whether the task is harvesting a crop or handling live broilers.

Another expanding field for engineers in agriculture is connected with animal feeding systems. Whether producing turkeys or milk, there are essential operations, either manual or mechanical. Where machines do the work less expensively than labor, or better, mechanization is the preferred alternative.

Engineers are now combining computers and electronics with machinery in agricultural operations. An example occurs in dairy feeding operations, where a sensing device identifies individual animals by ear tags and, with the animal's milk record in a computer, signals a tailored ration of feed. Mechanical feeding systems are increasing throughout agriculture—a result of the work of many professionals in designing and building specialized equipment.

With approximately 10 percent of U.S. cropland under irrigation, there are many career opportunities in this field. Irrigation engineers in California citrus orchards may focus on drip-irrigation installed tree by tree. Instead of using water by acre-feet, orchardists—through drip-irrigation methods—use water by the acre-inch. The result is a sharp increase in water-use efficiency, such as through reduced evaporation losses.

In Nebraska an irrigation specialist is likely to concentrate on developing pumps and nozzle orifices (openings) to improve water-volume delivery per unit of energy in pumping. Other engineers in this field may be working on gravity flow in ditch irrigation, with sediment catch-basins for this water at the low end of the row.

Engineering careers in agriculture also extend to environmental control systems. These systems may focus on misting systems in large dairy operations in the West or on environmental conditions such as humidity and temperature in hatcheries that mass produce baby chicks in Georgia.

In the field of environmental quality, there are many engineering opportunities. Some involve soil erosion control to stem sediment loss from fields and thus reduce sedimentation and eutrophication in lakes, reservoirs, or rivers. Others involve waste treatment—such as the concern with waste wash-water from vegetable and poultry processing plants—or in improved systems for storing and handling dairy and livestock wastes.

CROP INPUTS

Seed

Cytogenetics (the study of cells, chromosomes, and genetics) and cloning (form of asexual reproduction) are just two of the tools used by the commercial seed industry to improve the germplasm base for American agriculture. Genetically improved plants, which provide the basis for high crop yields, start with the seed.

In these seeds, plant breeders combine genes that not only impart disease and insect resistance but also boost nutritive value and tolerance to drought. Plant breeders also can manipulate maturity date, tolerance to soil acidity, palatability to livestock, and other important plant characteristics.

Thus, producing seed is a highly developed commercial business, whether for large acreage crops, such as corn, or for home gardens. The business provides career opportunities for plant breeders, pathologists, biochemists, and other specialists. Once genetically pure seed is available, commercial reproduction is accomplished by specialists or under their supervision.

Long-term commitments are essential in this industry. Development of a corn hybrid, typically, takes eight to ten years, according to the American Seed Trade Association. The germplasm base for a crop is under continual development by the seed industry as the make-up of insect and disease resistance shifts and as improvements in yield potentials are made.

Fertilizer

Of the many factors limiting crop yields, hardly any equals the extent or severity that results from plant nutrient deficiencies.

Plants require 16 essential elements, or nutrients, for growth. Even virgin soils are deficient in one or more of these in terms of providing enough nutrients for the high crop yields necessary for profitable agriculture. The function of commercial fertilizers is to provide these nutrients in forms readily available to plants.

Careers in this industry include those of mining engineers, chemists, and process engineers at potash and phosphate mines and processing plants and those involved with production of nitrogen fertilizers and micronutrients. Agronomists provide information on crop needs for nutrients and set conditions for tailoring fertilizer products to fit those needs.

Between production plant or mine and the local retailer there are many who have careers in the fertilizer industry in storing, transporting, and marketing. As fertilizers are large-volume bulk commodities, brokers and merchandisers are also needed to perform roles similar to those in the grain industry.

Also much like grain, basic fertilizer products—such as ammonia, urea, phosphates, and potash—are commodities extensively handled in international trade. Production of ammonia is tied closely to supplies of low-cost natural gas, such as in North America and the USSR, since gas provides an essential ammonia ingredient—hydrogen. It is the only fertilizer produced by a synthesis process, and from it most other nitrogen fertilizers are produced.

Phosphate and potash mines, however, are located where large natural deposits occur. The largest phosphate deposits are in the United States and Morocco, and the largest potash deposits are in Canada and the USSR.

As demand for fertilizer depends upon the magnitude of planted acreage and on general economic conditions of agriculture, there are career opportunities for economists and market researchers in both privately owned companies and cooperatives that produce fertilizers.

Production economics, where costs and benefits are closely evaluated, play an important role in fertilizer use, since they are one of the most expensive inputs purchased by farmers. Fertilizer costs per acre of corn, for instance, can be $50 or more; on some specialty crops such as tobacco, over $100. Sales representatives and service specialists in the fertilizer industry have an important role in providing information on the economic returns from this input.

Retailers provide the largest number of career opportunities in the fertilizer industry. These local businesses, numbering approximately 12,000, offer opportunities ranging from service personnel, who assist farmers with agronomic information, to accountants, who manage inventory turnover and accounts receivable. Many retailers operate a number of retail plant sites or outlets; thus, several plant managers may work for a single retail firm.

Fertilizer retailers can be privately owned businesses or cooperatives. Careers in the two areas, however, are similar, with emphasis on agronomic and business management backgrounds. Managers for both buy, store, and apply typically 5,000 to 50,000 tons of fertilizer annually. Retailers purchase supplies from basic producers, importers, brokers/traders, and wholesalers.

Careers at fertilizer retail centers often involve handling a broad range of farm supplies, such as lime and pesticides. Tonnage of lime sold at some retail operations is nearly equal to fertilizer tonnage, and much of the same application equipment is used.

Application for the customer of fertilizer, lime, and in some cases pesticides, is a major part of the services offered farmers by farm retailers. Such application is most efficiently done with large volume equipment that is impractical for many farmers to buy and own. As much of this involves highly specialized equipment, there is a sizable engineering design and construction industry that provides fertilizer plant and application equipment. A number of engineers have careers in this industry, producing handling, storage, and transportation equipment for liquified ammonia, fertilizer solutions, suspensions, and solid materials.

Another business which also offers many career opportunities is that of soil testing and plant analysis. These provide the farmer's best diagnostic means by which fertilizer use may be adjusted closely to needs of individual crops on specific soils.

Improved laboratory equipment and analytical methods for handling large numbers of soil and plant samples enable government and private laboratories to provide detailed information on degree of soil acidity, or alkalinity, and nutrient status. These laboratories provide careers for chemists in analytical work, as well as in interpreting test results. The latter is usually done with computer programs and integrates a large number of factors in making a fertilizer recommendation for an individual field.

A summary from the USDA showed that approximately three million soil samples are tested annually. The number is closely divided by government and commercial laboratories.

Although the number of soil samples tested annually has changed little in recent years, the number of plant samples analyzed has been increasing. Currently, this number is close to half a million annually, with commercial laboratories handling the majority of the tests.

Pesticides

Food and fiber crops are attacked by a host of pests. One account of these was made by the Monsanto Agricultural Products Company:

- Weeds—30,000 species, of which 1,800 cause serious losses
- Insects—10,000 plant-eating species
- Nematodes—3,000 species of such tiny soil-living worms, of which 1,000 cause serious damage.

To this list must be added viruses, bacteria, fungi, and rodents.

Until about World War II, there was little defense against these pests other than a hoe or a mechanical cultivator. According to the National Cotton Council of America, the first chemical weapon—the first insecticide—was provided to cotton producers in 1916. It was calcium arsenate. This remained the principal pesticide until World War II.

Even with our modern pesticides today, there are large crop losses to pests. In the case of cotton, for example, the National Cotton Council has estimated the following annual losses:

Cause	*Cotton loss in bales*
Weeds, such as Johnson grass	700,000
Diseases, such as verticillum wilt	1,500,000
Insects, such as bollworms	950,000

Since World War II, careers in chemistry, insect and plant physiology, and in chemical engineering have provided a wide array of chemicals for controlling plant and animal pests. More recently, biological means of protecting crops have joined the chemicals in pest combat.

Such products for agriculture have no equal in terms of development expense, development time, nor in extent of government regulation. Monsanto reported that the effort to take a single product from the discovery phase to the marketplace can involve a decade and cost millions of dollars. Additionally, tens of millions of dollars are required for a production facility to produce the new product.

Careers in pesticide development focus on research and testing. Specialists in testing must establish that the product works against the pest, while not damaging the crop or animal. Also, the product must not pose serious risks to humans, livestock, or the environment.

Lastly, a new pesticide cannot be used without governmental authorization. With the extensive role of government in controlling availability of pesticides, career opportunities exist for toxicologists, chemists, entomologists, regulatory agents, and other specialists.

Since 1910, pesticides have been under government regulation, administered by several different agencies.

Today, primary responsibility lies with the Environmental Protection Agency (EPA), which has administered the Federal Insecticide, Fungicide, and Rodenticide Act (FIFRA) since the EPA's formation in 1970. Even private citizens are subject to EPA regulations because use of a pesticide, inconsistent with the label, is a violation of law. And, for certain pesticides classified for "restricted use," only licensed applicators with special training are authorized.

The number of U.S. and international pesticide companies and products in existence gives an indication of the variety of careers that exist in private companies, as well as within the government. In a recent report the EPA listed the following for the U.S.:

- 30 major pesticide producers
- 100 companies marketing active ingredients
- 3,300 product formulators
- 29,000 distributors subject to EPA health and safety regulations.

With current public concern about pesticide use and the residual effects of these products on the environment, there is increasing interest in developing nonchemical control methods for insects, diseases, and other pests. Natural parasites are an example, as well as managerial or cultural practices, such as timing of application and handling of crop residues. Advancements in biotechnology also offer alternatives in limited cases.

Opportunities for careers in developing nonchemical pesticides will likely increase significantly in the future. Even use of predatory or parasitic insects is increasing. For example, in the cotton industry entomologists, insect physiologists, and others specializing in cotton have developed an integrated pest management (IPM) system.

In the 1950s career specialists with cotton insect control noticed that boll weevils were developing resistance to organochlorine insecticides. Subsequently, organophosphorus insecticides came into widespread use. But in the 1970s the tobacco budworm, a member of the Heliothis species, became resistant to these insecticides and posed a major threat to cotton production in Texas, just as the *Phytophthora infestans* was attacking potatoes in the 1990s.

To combat this resistance case, the EPA approved the use of synthetic pyrethroid insecticides. But now there are signs that the *Heliothis* species is developing resistance to these as well, especially late-season generations of the pest. Thus, an IPM program has been developed to avoid use of pyrethroid pesticides early in the season when predatory and parasitic insects can build up and control the cotton pests. When the effectiveness of these natural agents subsides, then pyrethroid insecticides are used.

This simplified case involving just one pest problem illustrates the variety of disciplines for those who choose careers in pest control. Chemistry of the pyrethroid and how it kills members of the *Heliothis* species, such as the budworm and boll weevil, is basic to these careers. Understanding the physiology and parasites of these worms also is critical. Toxicologists are those who determine just how toxic pyrethroid is to the insect, as well as to humans, to other forms of life, and to the environment.

These careers extend to the field. There, specialists measure insect growth population, recognize infestation, and provide essential services to the farmer, as well as to the sales personnel of pesticide producers. These specialists often are called "scouts." In effect, they are diagnosticians.

With such diagnostic information, the technical representative of a pesticide company or a registered pesticide applicator works with a local supplier of pesticides in recommending proper chemical application to the farmer. Or, the farmer may have sufficient training and experience with the pest to know the best alternative control.

Careers in commercial pesticide application may include a specialist at a fertilizer retail operation or a commercial aerial applicator covering up to 1,500 acres in one day. Again, specialization is required, especially with "low-volume" pesticides that can involve application of less than one pound of an active ingredient per acre. Such products may require micro-droplet sprays with specially designed nozzles.

Careers in this field of insecticide application are expanding as product use becomes more highly regulated, and as insecticide development becomes more demanding in terms of compatibility with other products, such as liquid fertilizer.

ANIMAL PRODUCTS

Careers in animal products span a wide range, from that of a chemist producing feed-grade calcium phosphates at a fertilizer plant to a sales representative of a pharmaceutical company that produces sulfadimethoxine as a drinking water additive to control coccidiosis, an intestinal infection of poultry. Such careers include ones for specialists in nutrition, parasitology, physiology, animal diseases, etc. There also are many careers related to sales and service, plus research and regulatory positions within government.

Similar to the situation with commercial crops, there are many diseases, viruses, parasites, and other pests that reduce milk production, rate of live-weight gain, or kill poultry and livestock in commercial operations.

Through biotechnology and mass production of growth substances, new careers have opened in industry and with government research agencies, where the objective is to stimulate normal rates of gain in milk and meat production. Some of these substances are growth hormones; some are anabolic agents—a substance that stimulates the pituitary gland to produce a growth hormone, such as somatotropin.

These developments also include the swine industry. One company, such as International Minerals and Chemical Corporation (IMC), may specialize in commercial production of a synthetic porcine somatotropin product for swine, while another, Monsanto, may specialize in bovine somatotropin to increase milk production in dairy cattle. Somatotrophin is a protein, and recent technology has made commercial production through biosynthesis feasible. When given to swine, this protein increases the rate of weight gain by converting feed more efficiently, and it produces more lean weight.

IMC provides an illustration of a company with specialized careers in this animal-product industry. In one IMC division, specialists produce an enteric intestinal antibiotic to increase weight gains in poultry. In its animal health group, IMC professionals produce zeranol, an anabolic agent. This product, implanted under an animal's skin behind the ear, stimulates production of somatotropin in beef animals. This, in turn, enhances protein deposition and promotes skeletal growth. The result is a 7- to 10-percent improvement in feed conversion and up to 40 pounds per head of increased feedlot weight gain in 100 days.

In still another division of IMC, the company produces a calcium phosphate feed supplement. Each of these nutrients, calcium and phosphorus, is required in large quantities for rapid livestock growth. High energy feed, such as grains, does not contain sufficient quantities of these two nutrients, so they are provided as feed supplements.

Industries that provide such animal products offer a number of career opportunities. Some will be in basic research, delving into chemistry with the most sophisticated laboratory equipment available. Others will be as bacteriologists, or specialists in animal nutrition and diseases; or as marketing, sales, and technical service representatives taking the product to the consumer.

The FDA, in a sense, is to this animal-products industry as the EPA is to the plant pesticide industry. Careers in the FDA are involved in testing chemical residues in processed meats and milk that may be left from use of growth stimulants, bacteriacides, or insecticides.

Another government agency, Animal and Plant Health Inspection Service (APHIS), is responsible for quarantine and eradication of animal diseases. As in the FDA, there are many careers within this agency of the USDA, from researchers to inspectors at airports and border stations.

CAREER OPPORTUNITIES

Agricultural input industries and related government programs provide opportunities for many careers, from engineers designing farm equipment to salespersons of medicated animal feeds. Examples of these careers are:

- Agronomist
- Animal health products salesperson
- Chemist
- Entomologist
- Equipment designer engineer
- Fertilizer broker/trader
- Fertilizer retail operation
- Fertilizer sales
- Field scout-fertilizer/pesticide retailer
- Irrigation engineer
- Market researcher, analyst
- Pesticide development
- Pesticide farm safety director (EPA)
- Pesticide registration (industry, EPA)
- Pest management
- Plant breeder
- Plant geneticist
- Plant pathologist
- Research technician
- Sales representative
- Service representative
- Soil test lab manager
- Soil test lab technician
- Toxicologist
- Waste management engineer

CHAPTER 6

SELLING WHAT THE FARMER PRODUCES

Probably the only valid definition of a business purpose is to create a customer. For agricultural marketing, this business purpose applies to customers in the local supermarket, as well as to buyers of U.S. agricultural exports around the world.

Collectively, there are many functions in agrimarketing. However complex, the basic function is to enable most of us to eat what we have not produced nor have had any hand in storing, transporting, or marketing. Combined, these functions account for a very large part of the economic size of the food and fiber sector, which makes up approximately 20 percent of the nation's gross national product (GNP).

It has been estimated that those who work in the agricultural marketing sector are four times more numerous than the farmers who produce the food. Grocery stores and supermarkets illustrate the magnitude of this marketing complex. Not only are annual supermarket sales in excess of $200 billion, but additional millions are spent on the storage and transportation of food products.

In 1992 over 100,000 managers worked for grocery stores and supermarkets and these businesses employed more than 900,000 cashiers, half of whom were on part-time schedules. These jobs can prove good entry-level positions for those interested in a retail career, and may eventually lead to management positions with more responsibilities.

Such numbers illustrate the size of the agribusiness industry. In this sector of agriculture, marketing skills, economic analysis, and business acumen come to the forefront for careers. Awareness of economic and international conditions is critical to commodity contracts with razor-thin margins. The ability to perceive consumer preference trends is a necessity for merchandising careers, whether one is specializing in frozen foods or in retortable pouches tailored for microwave ovens.

Training for careers in agrimarketing is heavily weighted towards market education. Knowledge of commodity markets, economic principles, trends in consumer purchasing power and preferences, financial management, inventory management for a high turnover, and extensive computer skills become the stock-in-trade for such careers. And, with large numbers of employees, personnel management is another large career field in agrimarketing.

INTERNATIONAL MARKETING NEEDS

International activities are particularly important in agrimarketing. Sales of processed broilers, livestock embryos for transplants, and large volumes of commodities, such as soybean and cotton, illustrate the range of careers in the international area. Large grain companies staff offices overseas with specialists to handle business on the spot and to get an early word on upcoming tenders (offers to buy) for agricultural products by foreign buyers, which may be governments or buying agents on behalf of governments. Obviously, they want career people with professional credentials, but language skills and knowledge of local governments and cultures also are critical. Some careers in agrimarketing require a degree in law, with specialization in commercial and international law.

Supporting the private sector in the international agrimarketing field is the diplomatic service of the USDA. This service consists of professional agriculturalists posted in about 70 American embassies worldwide.

In 1978, the Agricultural Trade Act established the diplomatic title of "counselor" for these professionals, who were previously known as "attachés." These counselors represent the U.S. secretary of agriculture—not the secretary of state—and handle all matters of agricultural trade information, food aid, and technical programs. An underlying objective of this group of counselors, administered by the USDA's Foreign Agricultural Service, is to promote U.S. agricultural exports and to maintain market outlets for a productive domestic agriculture.

STORAGE

The first step in marketing is selling the basic product, whether it's oranges in Florida or grain in Iowa. Next comes storage, especially important for the large volume grain commodities.

Sale at elevators, gins, or processing plants begins the process of establishing a commercial value for the product. The quality or grade is judged and, with this, commercial value is set. Agricultural products then begin their journey, often circuitous and indirect, to consumers.

Volume of storage is largest for grain commodities and provides a diverse range of careers in this industry, starting with the manager of the country elevator. Diversity even exists there, because such an elevator may be independently and privately owned. It may be owned by a farmers' cooperative or be a line elevator owned by a private investor grain company. Its capacity may be as small as 50,000 bushels or as large as a couple of million bushels. A typical country elevator has capacity of about 100,000 bushels, or about 3,000 tons—enough to fill 150 20-ton trucks.

Even selling grain at the country elevator is no longer a simple task. According to Cargill, the large grain conglomerate, there are at least seven ways a farmer may sell at the country elevator to meet cash flow, debt service, or special financial requirements:

1. Cash sale—spot sale that day, with immediate payment.
2. Cash sale/deferred payment—same as cash sale, but payment is deferred, to balance tax years, for example.
3. Cash sale/deferred delivery—a form of "forward pricing," where a cash market transaction arranges delivery later at a definite cash price.
4. Unpriced sale/basis contract—farmer and elevator sign a contract so farmer can sell later based on his judgment of the futures markets.
5. Option to purchase—contract gives the elevator the right to buy at a predetermined price and shipment period.
6. No price established (NPE)—contract on deferred sale of a specific quantity and quality of grain. No price agreement until later within limits of the contract.
7. Average pricing—sale will be based on average of 12 successive monthly pricings.

Not only does the farmer need to be an expert in the commodity market in making the best sale possible and in the best possible form, but the manager of the country elevator has to be a master at it.

Assisting the elevator manager, usually, is a specialist in grain sampling and grading, or the manager may do it personally. Career positions in grain grading, as in meat inspection, are critical to the whole agrimarketing network. Confidence in quality of product and uniformity of quality criteria among elevators and processing plants are essential to orderly, efficient marketing.

Grain Grading

In the U.S. grain industry there has been a uniform federal grain grading system since 1908, and the U.S. Grain Standards Act of 1916 provided the first uniform national standards by authority of law. However, under the U.S. Grain Standards Act of 1976, only grain for export must be officially inspected.

Hence, those with careers in grain marketing, such as with the USDA's Federal Grain Inspection Service (FGIS), quickly master the specifications for grain classes (such as white or yellow corn) and grades (quality characteristics such as moisture, test weight, and percent damaged kernels). Although only grain for export is officially inspected by USDA licensed inspectors, most grain in commercial transactions is graded according to uniform national standards.

For example, requirements of U.S. No. 1, hard red winter wheat are:

- 60 pounds per bushel—test weight
- 0.2 percent maximum heat-damaged kernels
- 2.0 percent maximum total damaged kernels
- 0.5 percent maximum foreign material
- 3.0 percent maximum shrunken, broken kernels

After establishing the quality of grain received, the elevator operator next focuses on protecting it. Back-up support for this comes from engineers who design grain handling and drying equipment and from specialists who specialize in insects and pathogens such as fungi.

Warm temperatures and moisture foster growth of the fungi that invade seed through cracked or broken seed coats. Such fungi not only reduce the quality of grain by fostering molds, but can also produce mycotoxins, compounds toxic to animals.

Many employees in the grain industry make careers of sustaining grain quality. These professionals range from specialists in grain chemistry to engineers responsible for air velocity. Different grains have optimum combinations of moisture and temperature for storage. Moisture content of soybeans, for example, should be below 12.0 to 12.5 percent, whereas 14.0 to 14.5 percent moisture is safe for corn in storage.

Other Opportunities

Similar career opportunities exist throughout the handling and storage industries for all major volume agricultural products. Fresh vegetables, fruits, processed meats, and even canned goods have unique storage requirements to prevent or to minimize loss of quality. Careers in these fields range from bacteriologists to metallurgists.

For example, engineers can control the storage atmosphere for apples so as to extend their freshness by five or six months. Such an atmosphere has 5 to 20 times less oxygen and 8 times the concentration of carbon dioxide as the air we breathe. Or, the metallurgist may develop additives to the inside coatings of metal cans to prevent acid products, such as tomatoes, from "detinning" the can while in storage.

Attention to storage quality, whether at the country elevator or at a beef warehouse, carries through to the consumer. Career professionals in this field, for example, know that corn damaged with air that is too hot will produce black spots in grits. Popcorn with too little moisture does not produce enough steam on heating to liquify the starch and to burst the kernel. Too much heat in drying corn modifies starch and oils in the seed so as to give a low yield of these constituents when the product is processed at "wet" corn mills. Improper drying of wheat can damage grain protein, cutting the production of gluten in baking and thus cutting the bread quality.

The consumer never sees these career professionals at the bread counter or in the meat section. But these behind-the-scene career professionals are essential to our supply of quality foods.

The continuing lines of storage in agrimarketing are illustrated by terminal elevators where managers are principal buyers from the country elevator. Generally, the terminal elevators are located at port cities, such as Norfolk or New Orleans, or at major river barge centers, such as St. Louis.

From these elevators, grain is sold to processors, millers, distillers, feed manufacturers, and exporters. Here commodity experts make careers out of buying, selling, and market hedging to keep the "river of grain" flowing. Also, there are career specialists in grain quality, grain conditioning, and grain blending.

Being a terminal merchandiser—either working independently or with a buyer such as a miller—is another career in agrimarketing. Grain merchandisers buy and sell cash commodities and perform the vital role of moving grain, ultimately, from producer to end users.

Through skills in making bids (proposals to buy) and offers (proposals to sell), merchandisers "move" grain, at least by ownership title, further along the marketing chain. In the process, merchandisers hold title to the grain during the period between purchase and sale.

In taking a position in the cash market—buying or selling—merchandisers deftly cut risks from adverse price fluctuations by "hedging." When buying cash wheat, for example, they will sell contracts for wheat futures. When the wheat is sold in the cash market, merchandisers will buy back these contracts. They hedge the commitments so that positions in the two basic commodity markets—cash and futures—are adequately balanced.

Another career opportunity in agrimarketing is to become a broker. They differ from merchandisers in that they do not take legal title to commodities. Instead, brokers function as agents and usually are paid brokerage fees by the seller. Their role, as in the case of the local real estate broker, is to bring buyers and sellers together, whether for cottonseed oil or for iced fish.

Careers in storage of fresh produce and frozen foods also involve working to preserve product quality.

Food chemists and other specialists monitor stored frozen foods for quality. In general, frozen foods have excellent retention of vitamins and nutrients below 0° F. Above 15° F, however, there is a loss of easily oxidizable vitamins.

TRANSPORTATION

Modern agrimarketing systems utilize all major modes of transportation—trucks, railcars, river barges, ocean vessels, and airplanes. Pipelines may be the only carrier not used in the agrimarketing system.

Obviously, transportation specialists play a major role in this system. They shop for, or devise, low-cost transportation services that also will assure maintenance of product quality. Transport of refrigerated beef from Kansas City to Charlotte, North Carolina, may involve shipment by rail to Atlanta and then transport by truck to Charlotte. Such a combination is called "intermodal" transport, since it utilizes two or more modes of transportation.

In the case of practically all grain shipments by truck, the sale is based on the amount and quality of grade, determined at the destination. Conversely, with transport by unit trains (when all cars within a given train are loaded with a single commodity), the price is based on the grade at point of origin, usually on an official grade set by a federally licensed inspector. And, for grain sold on a deferred shipment basis, the seller is responsible for the grain's quality to its destination.

Two acts by Congress in 1980 have largely removed the role of government in regulating rates that railroads and truckers charge shippers. The Staggers Act deregulated rate-making of all rail carriers, giving greater rate-setting flexibility on the basis of contracts with individual shippers. Likewise, under the Motor Carrier Act of 1980, rates were largely deregulated and greater freedom of entry given to truck lines to new areas.

Obviously, those with careers in agrimarketing responsible for transport need a thorough understanding of such laws, especially as they apply to grains. A specially designed commodity rate structure still applies to grain moved by rail. This rate structure, with its unique conditions, has no parallel in rates for other commodities.

The usual rail car for grain is a covered hopper car with a capacity of 100 tons. Both railroads and shippers own (or lease) fleets of these "commodity" cars. For special cars, however, such as refrigerated cars for vegetable or meat shipment, the burden of railcar ownership or lease is left to the shipper.

Transportation managers, whether for a terminal elevator or country elevator, recognize the cost disadvantages of single-car rates. Consequently, whether shipping soybeans from Jackson, Mississippi, or hard red winter wheat from Sioux Falls, South Dakota, transportation managers get lower per unit rail transport costs by using multiple cars. These shipments usually consist of 3, 5, 10, or 25 carloads.

In recent years, such multiple carload shipments have expanded to unit trains. In these, a single commodity is hauled by single trains made up of 25 to 125 cars.

Water Transport

Career professionals in river barge transport supervise movement of large volumes of grain, especially from terminal elevators to export points. Rates charged by barges have not been regulated and hence have been fully negotiable. Additionally, large grain companies own their own fleets of barges and towboats.

A number of individual barges can be joined together into a single "barge tow" to carry as much as 50,000 tons of Midwestern grain on the Mississippi to the Gulf for export. Such a barge tow (enough to fill two typical ocean freighters) carries the equivalent of five unit trains, each consisting of 10,000 tons (100 cars with 100 tons each).

Transport by water is usually the lowest cost mode of transport, and shippers of large tonnage commodities use it extensively.

Transportation managers face specialized arrangements for securing, or booking, ocean vessels. In many cases, the shipper owns no ships and, hence, depends on charters with shipping companies. Chartering is a highly specialized field involving knowledge of insurance, demurrage, and a host of other items subject to contract terms.

Except for those persons with careers in transporting agricultural products, few comprehend the volume involved. The case of exports illustrates the point. At a volume of 50 million tons of annual exports, there would be an equivalent of 2,000 ships required, each carrying 25,000 tons. Arranging for these, aside from the supporting domestic transport, is a large task for transportation managers and specialists in the agrimarketing field.

Large volume ocean freight is not limited to relatively low cost commodities, such as wheat and soybeans. More agricultural products are developing international markets. Hence, more efficient means of transport and storage of these are producing important career opportunities. A recent innovation in shipping containerized cotton to China is breaking a 300-year tradition of shipping the product in bales.

Shipment of orange juice from Brazil, reported by Cargill, is another example. In a pioneering move, Brazil has developed an ocean freighter to deliver bulk orange juice. This ship, with a refrigerated stainless steel tank, delivers 4.9 million liters (1.3 million gallons or approximately 5,200 tons) of frozen-concentrate orange juice. Obviously, when this ship unloads cargo in Amsterdam, there will be a few merchants in Western Europe in the market for containerized shipments of orange juice from Tampa, Florida.

The development of bulk, international shipments of orange juice illustrates the dynamic, innovative nature of the agrimarketing industry. It also illustrates the progress in distributing food from producers to consumers at increasing levels of efficiency and reduced costs. This progress, in turn, helps to increase market demand, an underlying objective in all marketing systems.

RETAILING

The modern supermarket is the target of the complex, orchestrated storage and transportation system. At this point of the funnel, people with careers in marketing display thousands of food products at arm's length for the consumer. Or, around the corner from the supermarket there likely is a fast-food business reflecting a rapidly growing prepared foods industry that also depends on agrimarketing. In the combination of these two industries lies the essence of preparing retail food products for the consumer.

In the supermarket, the consumer can find as many as 30,000 food and other consumer items—from boneless poultry to chef-designed, unfrozen, precooked, and microwave oven-ready products.

Not visible from the supermarket checkout counter are the supporting efforts of distributors, brokers, packers, and suppliers. People with careers in these areas are vital to successful retailing.

From these specialists, the career retail store manager obtains a number of "supports," including:

- Promotional support—local, national advertising.
- Quality, convenience, value—a quick-selling product.
- Expandable categories—growing product lines.

With the number of products rapidly expanding, there is an ongoing battle for supermarket shelf space. With this competition between products, the store manager constantly evaluates the bottom-line item—dollar sales per square foot per week. If hair-care products are giving sales of $15 per square foot per week, turnips yielding $10 will move over.

This sort of performance evaluation by the store manager, or sensory analyst, is made possible by such electronic devices as the UPC (universal product code) and the DPP (direct product profitability) analysis. With UPC devices, the manager can measure sales in a no-mistake system by using computer-driven scanners. DPP analysis calculates the relative profitability of items by including all known cost factors. Such are the modern tools for those with careers in retailing foods.

Competition among categories of products is also increasing. Frozen foods are battling with freezer-free products; plastic wraps are battling with glass; paper cartons claim better consumer education messages; longer shelf-stable products are displacing those with shorter shelf lives, and so on. Thus, food retailing includes many careers that involve computer expertise and proficiency in analyzing sales trends.

Another supermarket career field is labeling. Requirements of the Fair Packaging and Labeling Act, as well as the Federal Food, Drug, and Cosmetic Act, must be followed. Additionally, career specialists at the supermarket who concentrate on consumer education contribute to label format and content. Those specializing in labeling also will utilize new technology, such as carbon dioxide laser coding with special photochemical-sensitive labeling material.

Attention to packaging by those with careers in retailing is all-important at the consumer sales level. "Tamper evident" packaging has become more prevalent, and better materials such as shrink-wrap are now available. The latter improves consumer appeal and also increases the shelf life of products such as ice cream by cutting moisture loss. New "barrier plastics" that prevent passage of gases, especially oxygen, as well as moisture, offer convenience to customers and illustrate how much of our packaging technology is tied to plastics.

People with careers at the supermarket continue the role of those in food processing and on the distribution line by assuring wholesome consumer food products. Sanitation specialists, food chemists in supermarkets, continue the constant vigilance of protecting food quality. At the supermarket, they face numerous requirements of the Food and Drug Administration (FDA), from inspection of seafood products, to general cleanliness, to assuring temperatures below 0° F for frozen foods.

Continued diversification and expansion of retail operations to meet broadened consumer interests are also impacting food retailing. The result is an expansion of career opportunities in bakery management, deli operation, and even in cafeteria food services at supermarkets. Such instore services are providing prepared foods in competition with the fast-food outlet around the corner. The force that drives this is the desire to provide additional consumer convenience—at a profit.

Agricultural Marketing Service (AMS)

Additionally, there are many careers in the USDA's Agricultural Marketing Service (AMS). Specialists there facilitate orderly marketing of quality products through:

- Establishing standards for product grades;
- Administering marketing agreements, such as for fruits and milk;
- Providing marketing information;
- Enhancing food quality;
- Conducting research in marketing.

Marketing specialists in the AMS thereby contribute to our commercial system of agriculture in numerous ways, from setting quality standards on turkeys to providing 24-hour market information to buyers and sellers.

Commodity Futures

Careers in marketing make commercialization of agriculture possible. Without these careers at the country elevator or as a trader at the Chicago Board of Trade, movement of a product from the producer to the consumer would not occur within the U.S. system of commerce.

A complex system of commodity futures trading has developed that plays a critical role in agricultural marketing by spreading risk among buyers and sellers. Farmers, food processors, grain merchants, and speculators all participate in this system that moves nearly all agricultural commodities to market.

The basic unit in this trade system is the contract. For corn, for example, this basic trading unit is 5,000 bushels on the Chicago Board of Trade. Such contracts are sold, bought, hedged, and leveraged. These commodity market transactions involve the efforts of commodity brokers and analysts, as well as many others.

CAREER OPPORTUNITIES

Thus, from initial acceptance of a load of grain at the country elevator to sale of a loaf of bread at the checkout counter of a supermarket, there are opportunities for many careers in agricultural marketing industries. Examples are:

- Account Executive
- Accounting
- Agricultural counselor, attaché
- Agricultural establishment inspector
- Agricultural loan officer
- Cattle buyer
- Commodities broker (grain, cotton, etc.)
- Commodities analyst
- Commodity pool operator
- Commodity trade advisor
- Computer specialist
- Consumer information specialist
- Cotton gin manager
- Dairy processing equipment operator
- Economist
- Elevator operator
- Export financier
- Financial analyst
- Floor trader
- Food engineer
- Fruit/vegetable grader
- Grain buyer
- Grain inspector
- Grain merchandiser
- International specialist
- Packaging specialist
- Personnel manager
- Produce buyer
- Professional services, consultant
- Safety director
- Sanitation director
- Sensory analyst
- Trade specialist

CHAPTER 7

PROTECTING THE AGRICULTURAL ENVIRONMENT

The earth's ability to support life for an expanding population, which now exceeds five billion people, is possible only through efficient utilization of essential resources such as soil, water, and air. Likewise, the earth's ability to sustain increased resource utilization for larger populations will depend on the talents of many persons in agricultural careers devoted to protecting and conserving these resources.

Stewardship of these key resources for life—soil, water and air—is the essence of environmental protection. As each of these resources is considered renewable—to be used over and over—careers in environmental protection concentrate on preventing quality degradation that, in turn, would impair reuse of that resource.

Environmental protection also includes resource conservation, particularly for soil and water. Although considered renewable, quantities of soil and water are finite, and their conservation is a major element of environmental protection. Soil erosion losses from cultivated fields annually range from several tons to as high as 100 tons per acre. Yet, renewal rates through soil formation processes are very low—only 0.1 to 0.9 ton per acre per year.

SOIL CONSERVATION TEAMS

The environmental impact of soil loss through erosion includes the problem of soil sediment, which itself may become a pollutant along with the nutrients and chemicals it carries. Sediment increases turbidity of water and can choke river channels and reservoirs. It also is a carrier for much of the phosphorus and pesticides lost from fields.

The loss of soil through erosion—estimated to be four billion tons per year—and the resulting loss in land capability to produce agricultural crops were the initial reasons for soil conservation programs, but the environmental aspects are perhaps paramount today. Additionally, the cost of erosion after soil has been lost from the farm approaches $6 billion per year in terms of stream sedimentation, pollution treatment, and ruination of reservoirs.

Teams of specialists who address these concerns include those in a variety of careers. These include those who work with soil and water conservation districts, the Soil Conservation Service of the USDA, and regional and federal offices of the EPA, as well as those at state agencies.

Soil and water conservation districts are local government units organized under state law. Many follow multi-county borders and number nearly 3,000 in the United States. The primary role of specialists who work in district offices is to provide sound resource management information to land users and government units in an effort to help cut the heavy loss of the nation's valuable soils.

Managers and technicians with these districts work closely with the U.S. Soil Conservation Service, the principal source of federal assistance. At the national level, district personnel are represented by the National Association of Conservation Districts (NACD), a nongovernmental, nonprofit organization.

Several federal government programs offer careers in soil conservation that relate closely to environmental protection. Under the Conservation Reserve program, the USDA contracts with farmers to retire highly erodible cropland from production under 10-year agreements. A key requirement of this program is that farmers must keep this land under permanent cover, such as with grasses or trees, to prevent erosion. In return, the USDA pays annual "rent" to the landowner.

A relatively new provision in federal farm programs is termed cross-compliance. Under the 1985 Farm Bill, conservation activities are linked to farm program benefits. For example, it required those farmers with highly erodible fields to have an approved conservation plan in effect by January 1, 1990. Farmers without this cross-compliance were ineligible for any farm program benefits.

The Soil Conservation Service and the Agricultural Stabilization and Conservation Service are the two USDA agencies that manage such programs. They have personnel at county, state, and national levels serving in careers in environmental protection that specialize in soil and water conservation. Federal requirements for conservation plans at the farm level increase the opportunities for those in such careers.

Some in these careers focus on the technical aspects of soil erosion, such as quantities of sediment lost depending on slope, water velocity, soil type, conservation practices used, and other factors. Others focus on the chemistry of erosion as related to environmental quality and provide information on organic carbon levels, nutrient levels in solution versus nutrients attached to sediment, and on pesticide concentration.

WATER QUALITY

Water is nature's solvent. The oceans attest to this fact with a content of 35,000 parts per million (ppm), or 3.5 percent, salt.

Soil water, bathing plant root surfaces, is a dilute nutrient solution. Even in pristine streams, water is a solvent for solutes and minute silt particles. Indeed, where there is a natural solvent such as water, there will be solutes, dissolved substances. In nature, there is no pure water, not even in rainfall.

Surface Water

As a carrier, water likely surpasses wind in transport of solids. Alluvial plains and deltas, as well as sediment deposits in reservoirs, are results of water carriage.

Realistic environmental protection of water deals with tolerance levels of solutes relative to human health and environmental quality. Maintaining the level of nutrients and other solutes at acceptable ranges for human health and environmental quality is the objective of environmental protection. Environmental protection careers in agriculture involve aspects of both ground and surface water.

Career professionals include soil physicists, hydrologists, engineers, soil chemists, biologists, and environmental health specialists. Also, with emphasis on best management practices (BMPs), an increasing number of communication specialists are joining technicians in preparing information for farmers on BMPs for local soil/cropping systems.

As nutrient loadings to surface waters are reduced by "point source" control (control at industrial facilities), non-point sources such as agriculture and forestry are becoming the largest sources of nutrients that enter surface and subsurface waters. There are some estimates that more than 80 percent of the nitrogen and phosphorus entering streams come from agricultural land runoff. A large portion of pesticides entering surface water has the same origin, with the largest fraction usually adsorbed or associated with soil particles or organic matter.

Developing BMP's for different cropping systems and promoting their adoption by farmers is a major role of many in environmental protection in agriculture. Grass waterways, or filter strips, may be the best BMP for one set of circumstances, whereas catch basins, or impoundments, may be the best BMP for irrigation return flow conditions. Grass waterways may be the BMP most effective for catching sediment and thereby cutting phosphorus and pesticide loadings.

Such are the examples of practices tailored by conservationists and others working with different systems of soils, climate, and crops. The purpose of these practices is to reduce the addition of potential pollutants, such as sediment, nutrients, and pesticides from our food production system of crops, to the environment.

Ground Water

The NACD has reported that ground water is the most abundant freshwater resource in the United States. More than 20 times the combined volume of fresh water in surface lakes, rivers, streams, and reservoirs lies beneath the surface of the earth as ground water.

Hydrologists and soil physicists, many with the U.S. Geological Survey, explain that ground water fills the voids of porous earth materials, creating saturated zones that yield sufficient volume for wells and springs. It also makes up the large bodies of saturated sand and gravel comprising most aquifers, where the porous nature of solids permits rapid transmission of water. These career specialists also study transport time of "recharge" water. In some soils, it can take many years for portions of surface water to be transported from the plow layer of surface soil to deep ground water levels.

Except for injection, protection of ground water focuses on the downward movement of water that makes up recharge water. This water becomes the carrier of many minerals and salts and also can carry soluble pollutants to the ground water deposits.

Protecting ground water from nitrate and pesticide contamination is a priority objective in environmental programs at both the state and federal levels. This emphasis is appropriate in view of the fact that half of the U.S. population and 90 percent of rural residents obtain water from ground water. Levels of phosphorus and other plant nutrients in recharge water do not pose a problem to ground water because of their affinity to attach and remain with soil particles. Thus, protection of ground water from agriculture centers on concerns regarding nitrates and pesticides.

While environmental specialists and chemists measure nitrates in parts per million (ppm), pesticides are measured in parts per billion (ppb), or even ppb fractions, Ground-water surveys by the state and federal agencies show the presence in ground water of a number of pesticides used in agriculture. Several of the first to be identified were aldicarb in New York, EDB (ethylene dibromide) in Florida, and DBCP (dibromochloropropane) in California.

PUBLIC HEALTH CAREERS

Careers in the U.S. Public Health Service also are involved with ground water quality. This service has set a maximum of 10 ppm of nitrate nitrogen (45 ppm nitrate) as a safe level for drinking water. Health experts with the Public Health Service work closely with the U.S. Geological Survey in assessing nitrate levels in ground water.

Also, many agronomists work on the issue of nitrates in ground water. Nitrogen absorbed by plants is intercepted from water passing through soil, and hence, attention is placed on increasing crop-use efficiency of nitrogen to cut losses of this element to the environment, including those to ground water. Those in these careers include researchers, in both private industry and public agencies, who develop management practices and chemical treatments to increase efficient plant uptake of applied nitrogen.

Among the nitrate sources in ground water and in surface waters, the most easily controlled is fertilizer. Nitrogen contained in precipitation and plant sources, such as legumes, is not easily controlled. As nitrate levels in ground water increase, there will be added pressure on agronomists to improve significantly the recovery of nitrogen applied in fertilizers by crops to which they are applied. Typically, this recovery is presently 40 to 60 percent.

Nitrogen applied annually in commercial fertilizers amounts to 10 to 12 million tons. Nearly two-thirds of this is applied to two crops—corn and wheat. Thus, many agronomists who work with these crops either conduct research on techniques to improve plant uptake of nitrogen or work in educational programs promoting the use of practices known to cut environmental losses, such as proper fertilizer and application methods.

Those with careers in handling animal wastes also deal with nitrates in ground water. Wastes from feedlot and dairy operations result in high nitrate concentrations and thus pose significant localized sources of nitrates for ground water. Engineers working with these sites develop lagoons to hold animal manures, sometimes with special liners and containerized systems to reduce nitrate losses to ground water from leaching of such wastes.

NEW CAREER OPPORTUNITIES

As with most agricultural career fields, there are new opportunities in water conservation and environmental protection involving computer specialization. For instance, the increasing volumes of data on weather, chemical and physical characteristics of soils, aquifer profiles, fertilizer and pesticide use, and crop rooting patterns can be digitally incorporated into computerized systems.

An example of such a system, developed by Michigan State University, is the geographic information system (GIS). Working with this system, specialists incorporate details, such as conventional versus minimum tillage and solubility of specific pesticides, into site-specific data. Thus, a systematic analysis is provided for a given locality, which would otherwise be impossible, for use by individual geologists, soil scientists, pesticide scientists, and others working on ground water protection. By computer, digital layers of information can be combined to identify specific areas of potential ground water contamination from agriculture.

PESTICIDES AND ENVIRONMENTAL QUALITY

There are many careers that involve the development, marketing, and use of pesticides. The range is wide—from professional lawn- and garden-service representatives, who recommend a fungicide to control mildew on roses, to the chemist who develops information for a new pesticide registration application.

The "poundage" use of pesticides may have peaked in the United States in 1981 with 1.5 billion pounds (active ingredients), nearly three-fourths of which was involved in agricultural use. Since then, the volume has declined, partly because of crop acreage reduction, but also because of increased use of "low-volume" pesticides. An example is pyrethroid insecticides, a relatively new group of compounds, which may be applied at less than one pound per acre.

Of the pesticides used on major field crops, approximately 80 percent are herbicides. Insecticides make up 13 percent, and the remainder includes defoliants, fumigants, fungicides, and miticides.

One cropping practice that has increased herbicide use is minimum tillage. The success of this practice—which offers significant benefits in reducing erosion, in conserving energy, and in producing higher crop yields—depends on control of weeds. Minimum tillage, with little or no mechanical weed control, must rely heavily on chemical weed control.

Careers in the pesticide industry are concentrated within three primary groups: basic manufacturers, formulators, and distributors.

In each of these groups, entomologists, toxicologists, product registration specialists, and others spend much of their careers with government regulations designed to protect the environment.

Product Registration

Specialists whose careers involve research at the basic manufacturer level to meet requirements of product registration are some of the first in the long line of those who make pesticides available commercially. Providing for environmental protection is one of their major responsibilities. They must obtain from the EPA an approval registration and approved label for each pesticide. Both are requirements that must be met before a pesticide can be sold in the United States.

Product registration, a premarket review and licensing program under the FIFRA, requires extensive field testing in addition to laboratory and greenhouse research. Much of this work addresses questions of human safety and environmental impact. In addition to the many careers in private companies that develop data for human and environmental safety, there are scientists and staff personnel in the EPA administering the FIFRA.

Obviously the number of applications received in the EPA's Office of Pesticide Programs varies from year to year. Those men and women making careers there have handled as many as 15,000 applications during a year of which only some 15 new chemicals received registration approval. Most approved registrations are for new formulations containing active ingredients which have already been registered. Thanks to many competing products on the market, there recently were as many as 1,400 active ingredients registered for use in some 45,000 pesticide products in the United States.

Data on "environmental fate"—how a pesticide behaves in the environment—make up much of the registration documentation. With this information, EPA specialists can determine whether the product poses a threat to the environment. The EPA may classify a product as a restricted pesticide if toxicity data warrant. Such pesticides may be applied only by, or under, supervision of certified pesticide applicators, another career opportunity for many interested in the distribution and sale of pesticides.

The most important properties that determine whether a pesticide poses a threat to ground water are persistence and mobility. A product that is both persistent and water soluble has the greatest potential for environmental harm.

Although a number of pesticides are being detected in ground water, and especially in shallow wells in highly permeable soils, pesticide losses in ground water and surface water are quite low compared with nitrates.

However, environmental scientists are finding cases of increased pesticide residues in soils, which in time can lead to higher levels in ground water. Also, some pesticides, such as atrazine, are persistent in ground water year-round, but in low concentrations.

Most careers involved with the pesticide registration process, which meticulously seeks to protect the environment, are deeply rooted in the sciences of soil chemistry, toxicology, plant and insect physiology, entomology, environmental chemistry, and analytical chemistry. Clearly, such careers are research oriented, whether in private industry or with government agencies. Most require education in the sciences.

Label Approval

Concurrent with product registration is label approval. Preparation of the proposed label is the responsibility of the applicant. Again, specialists prepare the label information for pesticides. They provide explicit information on application methods and rates and instructions for use. One section of the label for many pesticides is "Environmental Hazards," which describes critical characteristics of the product, such as toxicity to aquatic invertebrates, and gives directions for avoiding contamination of water supplies.

Most pesticide industry careers described thus far are with the approximately 30 basic manufacturers and producers. These firms produce the active ingredients, the basic building blocks, for approximately 3,300 pesticide formulators. Many careers in this latter group have functions similar to those with basic producers. However, there are additional formulator careers that specialize in development of diluting agents, inert carriers, sticking agents, emulsifiers, and the like.

In addition, there are literally thousands of pesticide distributors in the United States. Each of the primary groups has specialists in education and training, but the distributor, working directly with users, has the greatest opportunity to provide information on environmental protection to the consumer. Proper use, according to the label, assures minimum risk to the environment.

Careers in the pesticide industry increasingly focus on efforts and talents needed to develop and use pesticides so as to assure minimum losses to the environment and minimum levels in the food chain. As long as pesticides are used as an essential input to agriculture production, losses to the environment cannot be totally eliminated.

The task of many career specialists within the area of agricultural chemicals, such as pesticides, continues to be development and use of crafted chemicals that provide selected benefits with no negative effects on the environment or on human health. Currently, few pesticides may be totally able to pass such a test. Therefore, benefits and possible disadvantages associated with some losses to the environment must be constantly weighed.

CAREER OPPORTUNITIES IN ENVIRONMENTAL PROTECTION

Although divergent and varied, examples of career opportunities in environmental protection are:

- Agricultural engineer
- Agronomist
- Analytical chemist
- Biologist
- Certified pesticide applicator
- Computer specialist
- Entomologist
- Environmental health specialist
- Farm conservation planner
- Geologist
- Hydrologist
- Insect physiologist
- Microbiologist
- Parks and recreation administrator
- Plant physiologist
- Public health official
- Soil chemist
- Soil conservation district manager
- Soil conservation district technician
- Soil conservationist
- Soil physicist
- Toxicologist

CHAPTER 8

RESEARCH TO HELP THE FARMER

Those who like to explore, experiment, and understand, and those who are fascinated with living things will find careers in agricultural research and development highly rewarding. They will be the ones hastening development of the biological age.

Researchers seek to explain why things take place; to establish cause and effect—mode of action. With this knowledge one is more capable of making things happen, whether attempting to tailor a chemical to interrupt a unique enzyme in an insect, or whether manipulating forms of food starches in wheat flour to keep blood glucose and insulin levels in check.

In contrast to the more inquisitive nature of research careers, those in development usually concentrate on converting ideas into reality. In many respects, research and development are inseparable, and the distinction just given does not fully apply to many who have careers in research and development. The two generally go hand-in-hand for a career.

Careers in research and development offer some of the most rapid growth prospects of any in agriculture, whether in production, product conversion, or processing. Individuals in these fields utilize the most modern scientific equipment available. For example, the soil chemist will use the atomic mass spectrometer to measure movement of isotopic-labeled nitrogen in soil water of a corn field. A biochemist may use high performance liquid chromatography to purify an enzyme in cell membranes that control passage of incoming and outgoing molecules. With such tools, and with the knowledge accumulated within the many agricultural sciences, those working in research and development are on the one remaining frontier in agriculture.

Those with careers in research and development also contribute to the existing body of information. By expanding our understanding of nature, these R & D specialists contribute to our ability to control or modify some of its parts as a means to increase food and fiber production. People with careers on the leading edge of this pioneering work specialize in narrow interest fields, such as biochemistry, genetic engineering, molecular biology, embryo chemistry, parasitology, and food chemistry.

Opportunities for such careers exist in numerous public organizations, from state universities to a variety of federally funded organizations. Industry also offers opportunities for many careers in research and development, especially those that produce proprietary products.

As expensive as research is today, it obviously is located where the money is, either from government funds or corporate earnings. Some industries with proprietary products spend as much as 5 to 10 percent of their sales in funding research. Others producing commodity products, such as the fertilizer industry, spend as little as 0.2 percent of their sales for research.

Whatever the narrow field of specialization may be for careers in agricultural research and development, they involve production of green plants (or vegetation for aquatic life), conversion of plants through poultry and livestock, or direct processing of plants. Thus, one logical way of examining careers in agricultural research and development is to review examples within these basic fields.

PLANT PRODUCTION

Photosynthesis, either in green leaves or in certain algae, is the basic production process for food. Even with a high-yielding corn crop, photosynthesis is notoriously inefficient—on the order of 1 or 2 percent—in converting the energy of sunlight, and water and nutrients from the soil, into food energy, that is, carbohydrates and protein.

Production of green plants as our primary means of photosynthesis is plagued with numerous natural adversaries. Nutrient concentrations in the soil solution under natural conditions are insufficient for optimum growth. Predators, such as nematodes, chew the roots; fungi clog the translocation channels within stalks; competing weeds consume precious water; and budworms may devour fruit at the top of the plant. Thus, many people in research and development careers in agriculture concentrate on producing higher crop yields through genetics, soil fertility, or crop chemicals in protecting the plant.

Germplasm, Seed

No group of packaging experts, nor even of molecular biologists, can match nature in the ability to package a miniature plant, with all its hereditary characters, into a single small seed. Geneticists build their careers around unraveling and recombining these hereditary characters—manipulating genes to increase plant leaf surface, enlarge the root system, increase the protein content of the grain, or increase insect resistance.

Relatively new career specialists are now joining geneticists. They are molecular biologists—cytologists who study cells—and genetic engineers. Each of these works on genes, those minute packages of DNA (deoxyribonucleic acid) that provide the genetic codes for all traits in every living organism. Whether in plants or animals, genes are located in each cell nucleus. They are units of a long DNA molecule consisting of four nucleotide bases—thymine, adenine, guanine, and cytosine. Genes are the codes of life and control the order in which some 20 amino acids are assembled to produce different proteins. Each gene encodes a certain protein.

Our chief benefactors of this knowledge are James Watson, of the United States, and Francis Crick, of England. Working together in 1954, they put the finishing touches on a model of a DNA molecule that they were building and then reportedly went to lunch one day telling each other that "a structure this pretty just had to exist." Today, this marvel is depicted as a thin, twisted ladder with the four nucleotide bases forming the crosspieces between the two long sugar-phosphate strands that make up the sidepieces.

Plant breeders, genetic engineers, biotechnologists, geneticists, and molecular biologists are different names for very similar careers in the study and manipulation of genes to modify codes to life traits. All living organisms, from bacteria to humans, share this same genetic code—a life of astonishing similarity, yet one of incredible diversity from the same building blocks.

A simple bacterium, consisting of only one cell, has 3,000 to 4,000 genes. No one knows the exact number in plants, but estimates are that there are at least 50,000 structural genes in each cell. These are organized into chromosomes, somewhat like beads on a string. While a cotton plant has 52 such strings, or chromosomes, bacterial cells have but one.

In organisms, such as plants, which reproduce sexually, there are two copies of each chromosome, each parent contributing one set to the offspring. This process is the mechanism by which the plant breeder secures new traits and introduces them into a new crop variety. Seed has been the traditional package.

Today, gene specialists and cytologists are turning to tissue culture and gene transplants to improve the efficiency of variety creation. Tissue culture enables these specialists to screen a large population of cells, quickly identifying mutations or variants. Monsanto has used this technique to select alfalfa plants from such a culture that exhibit resistance to a weed-killing herbicide.

The main thesis in molecular biology is that genes give the code for messages, which, in turn, produce a polymer—a protein—assembled from the same 20 amino acids. Hence, "gene engineering" is the process of identifying the gene responsible for a certain trait and moving that gene from one organism to another, a process called transformation. The DNA is cut and spliced and, in its recombined state, is called "recombinant" DNA.

In this process, we have the confluence of careers in molecular biology, genetics, biochemistry, and physiology. The process involves handling individual molecules; hence, the term molecular biology. Ohio State University has illustrated the intricacy of this process by comparing the length of one molecule of DNA to one kilometer. At this length, a single gene for engineering, cutting, and splicing would have a representative length of only one millimeter. Today, these specialists have a "gene machine," or DNA synthesizer, to assist them.

Benefits

Careers in this field of technology already have provided benefits to the grower and to the food consumer. One example of this progress in genetics is a variety of lettuce developed to be resistant to a fungus in California. Prior to this resistant variety, there was no practical control of the fungus that fed on roots of lettuce and, in the process, cut yield and spoiled flavor.

In the food chemistry area, wheat genes are being identified to determine whether the starch formed is of the straight chain variety (amylose) or branched chain (amylopectin). Dietary studies show that pastries with amylose are better than those with amylopectin in keeping blood glucose and insulin levels in check.

Thus, even the type of starch in crackers is subject to tools of genetic engineering.

However genes are spliced or combined, commercial production of most crops begins with seed. "Tissue culture" is still used for potatoes, as well as for grafting many commercial orchard fruit varieties. Once the plant breeder has produced the seed, "plant protection" careers step in.

As soon as it is placed in soil, a seed begins a battle to survive. Fungi, universally present in soil, are copious producers of enzymes that hydrolyze (decompose by combining with water) cellulose, pectin, and other constituents of seed coats and seed parts. Other fungi produce substances which are toxic to germinating seed. Soil conditions such as moisture and warm temperatures that favor seed germination also favor fungi growth.

Plant pathologists study these fungi, the enzymes they produce, and their mode of entry into plant cells. With this information, these specialists can develop compounds to combat such fungi. These career specialists develop fungicides for seed treatment as a means of controlling seedling diseases caused by fungi, such as *Rhizoctonia* and *Pythium.*

Such chemical protection helps to assure a vigorous, healthy plant from each seed. This protection is widely used in agriculture. Essentially all commercial seed of major crops, such as corn, is treated with a fungicide. In many cases, depending on the crop and locality, an insecticide is included in seed treatment.

With the large investment required to develop a new crop variety, there are increasing opportunities for careers in the seed industry to handle germplasm patents. Since 1970, when the Plant Variety Protection Act was enacted, commercial seed companies have had protection from "pirating" of germplasm for commercial purposes. Seed protection applies only to commercial sales, not to the use of a patented variety in research and development.

There also are careers in the USDA's National Plant Germplasm System. There, a bank of plant material germplasm is maintained, all of which is nonpatented and available to private companies for research and development.

PROTECTION AND REGULATION OF PLANT GROWTH

As described in Chapter 5 on careers in input industries, the crop protection, or pesticide product industry, is highly dependent on research and development. This industry, particularly among the basic producers, provides a large number of research career opportunities. Most of these companies provide products for the international as well as the domestic market, where competition in technology from such world giants as Imperial Chemicals Industries of England, Bayer of Germany, and Sumitomo Chemical of Japan is intense.

The need for long-term commitments to research and development by such producers is understandable in light of the time required to develop and commercialize a plant-protection chemical. Such a product, according to the EPA, may take six to nine years or more from laboratory through full EPA registration and eventual marketing to application. During this period, two to three years are required just for EPA registration approval, costing the applicant $2.4 million to $4.0 million for a major new ingredient. Total producer developmental cost for such a chemical is $50 million to $70 million. To this must be added the cost of the production plant or plants, which may be $100 million or more, according to the National Agricultural Chemicals Association.

Research careers in this field concentrate on mode of action. With an increasing body of information on plant physiology, plant cells, insect reproduction, and enzymes, mode of action can be anticipated for a given chemical with plants, animals, insects, and the environment.

In addition to research on mode of action, new plant protection chemicals require extensive testing for many other factors, including soil-water solubility. Researchers with careers in such programs use all the modern tools of chemistry, such as radiolabeled active ingredients.

With such isotopic labels, researchers can measure rate of decomposition to carbon dioxide, as well as extent of reaction with soil constituents and soil-water movements.

In the quest for products to protect selective plants, researchers add to the understanding of plants. For example, a Purdue University study of the fungus that caused the corn blight of 1972 uncovered new information about phenols, such as tannins. These phenols, which may make certain varieties bitter to livestock or which account for poison ivy-producing blisters, bind to protein molecules and change the way that proteins perform. In sorghum, these phenols attach to proteins as the grain is chewed and make the proteins useless as livestock feed.

A researcher who was studying fungal spores that infect corn leaves helped explain why many people in developing countries who live on excessive phenol sorghum often are deficient in protein.

Growth Regulators

Increased understanding of plant physiology leads to development of substances that can help regulate plant growth. An example is a plant growth regulator from Monsanto that is used on sugar cane to increase sucrose content.

Another example is the group of defoliants for cotton, a plant which has an indeterminant growth unless stopped by low temperature or chemicals. The use of chemicals is essential in order to harvest high yields of quality cotton. Defoliants are growth regulators, in a sense, as they initiate growth of an abscission layer of cells at the base of the leaf stem, causing the leaf to drop. Proper maturity and drying of the bolls are possible only on defoliated plants.

Development of such products represents tremendous capital investment, as well as vast amounts of proprietary information. To protect both, the basic manufacturer turns to U.S. patent rights. With extensive international trade in these products, where some countries recognize U.S. rights and others do not, there is an increasing number of people who deal with plant protection for plant protection and regulator products.

Under U.S. law, patent rights enable the patentee to exclude others from commercial exploitation of the invention for a period of 17 years. As development of a pesticide, including registration approval by the EPA, may take six to nine years, there may be only eight to eleven years of commercial-life protection for such a product. Specialists in this area, serving agricultural input industries, recognize that U.S. patent law makes no discrimination with respect to the citizenship of the inventor. They also are aware that the 17-year protection period applies only to commercial sale—not to research and development by another party.

OPPORTUNITIES IN NITROGEN CHEMISTRY

The varied tasks of converting nitrogen into protein offer a number of opportunities for careers in molecular biology and chemistry. Daily requirements for protein are on the order of 70 to 80 grams per person. Protein, typically, is 16 percent nitrogen, so the task of producing this daily "ration" involves 11 to 13 grams of nitrogen per person per day.

Microbiologists and soil chemists understand the elusive nature of nitrogen—largely a result of its multiple forms. In the elemental state, nitrogen is an inert gas, making up nearly 80 percent of the air we breathe. Combined with hydrogen, in one of its reduced chemical states, it is ammonia (NH_3)—a gas unless liquefied under pressure or refrigeration. Combined with hydrogen at a different valence, it is amide (NH_2) nitrogen, the form in life-giving amino acids and proteins. In its highly oxidized state, nitrate (NO_3) nitrogen is highly soluble in water and is the form that may pose health problems in ground water at certain concentrations.

Nitrogen Fixation

New career opportunities have developed in nitrogen fixation, especially in molecular biology. As described in a report by the National Science Foundation, specialists in molecular genetics are deciphering the genes that permit, first, the symbiotic relationship between a simple bacterium and a legume root to fix nitrogen and, secondly, the process of using energy from carbohydrates supplied by the host plant to fix—or convert—nitrogen in the soil air to ammonium nitrogen. It is this fixed nitrogen that ends up as protein in plants and that helps supply the 70 to 80 grams of protein that humans consume daily.

The goal of those who work with nitrogen fixation is to increase the quantity available for the plant's protein production. Nitrogen, or protein, deficiency is still a major plague for millions of people. Scientists estimate that global biological nitrogen fixation is 175 million metric tons, slightly more than half of which occurs within agricultural soils.

In addition to the 90 to 100 million metric tons of nitrogen added to the soils by nature annually, humans add some 70 million through commercial fertilizers, such as ammonia and urea. Yet protein deficit diets remain in much of the world.

The opportunities for careers to unravel the secrets of the nitrogen fixation processes and to improve these processes through gene splicing, chemical modification, or other means, is wide open. Such careers will concentrate on enzyme chemistry, for it is known that the enzyme nitrogenase is the "spark plug" in *Rhizobia* that splits the strong double bond of nitrogen. Once this bond is split, other enzymes come into play in pairing up the "hot potatoes" of reduced nitrogen and hydrogen to form amide nitrogen.

Those with careers in molecular biology will eventually unravel the mysteries of the symbiotic arrangement between bacteria and plants, as well as the processes that are tucked away in enzymes used by these bacteria to fix nitrogen. Once these secrets are known, careers in nitrogen chemistry will significantly alter how humans will secure daily protein needs.

Nitrosomonas

When nitrogen additions are made to the food production system of soils and plants, there are nitrogen losses to the environment. Here again, those with education in bacteriology and chemistry are finding challenging careers developing means by which another group of bacteria that oxidize ammonium nitrogen into nitrates may be manipulated. These bacteria, *nitrosomonas,* obtain energy from this oxidation process and are prevalent in all soils. Their activity completes another step in nature's nitrogen cycle.

In the nitrate form, nitrogen is completely mobile in water. Hence, water moving through soils (leaching) decreases nitrogen availability to plants by nitrate removal and adds to nitrate levels in ground water, underground aquifers, or in subsurface flows to lakes and streams.

Persons with careers in this field, such as specialists at the National Fertilizer Development Center, are discovering compounds, including formaldehyde and elemental sulfur, that inhibit urease. Thus, there are career opportunities to develop low-cost, effective means of controlled-release nitrogen fertilizers from urea by those who specialize in nitrogen and enzyme chemistry.

Other career researchers in bacterial physiology and chemistry help develop means for blocking the oxidative process of *Nitrosomonas,* such as bactericides. Nitrapyrin, a compound developed by Dr. Cleve Goring of Dow Chemical and marked as N-Serve, is an example. It is called a nitrification inhibitor and functions by killing the nitrifying bacteria within a given area of soil treated with the bactericide.

RESEARCH PUTS THE PIECES TOGETHER

The brief description of careers in research and development cited in this section on plant production serves only as an example of the many that exist in the biological sciences. With any of these specialties, however, the ultimate task is for researchers to put as many pieces of technology together as possible for the farmer. These researchers, in effect, create careers as transfer agents, taking science and technology from the laboratory and greenhouse to the field.

There are many researchers in state universities and federal research organizations who devote their careers to this application of technology. There also are many in industry, such as those with commercial fertilizer suppliers, animal feed-supplement manufacturers, and pesticide producers.

Some industries that serve agriculture provide careers in technology transfer through industry-sponsored associations. The Potash and Phosphate Institute, serving the fertilizer industry, is an example. Through its staff of professional agronomists and the research that it sponsors with state universities, this institute concentrates on fostering application of soil fertility research in the field to increasing crop yields and efficiency.

Through this industry-supported program, additional careers have grown. In effect, researchers have combined technology from a number of career specialists in crop production. The result has been a near tripling of average yields for major crops. Career researchers in this program have shown how to produce per-acre yields of more than 300 bushels of corn, 100 bushels of soybeans, and 150 bushels of wheat. Through such high yields, farmers are able to cut their production costs per bushel, thereby increasing profits and utilizing resources more efficiently.

NEW CROPS OFFER OPPORTUNITIES

As surpluses of food crops or resources for their production continue, American farmers may turn to alternative crops currently under development by botanists and plant breeders—provided these crops are profitable. For ages, the main focus on crop production, aside from cotton, has been for food. This focus may change, however, as crops that can be processed into industrial materials become economic alternatives for farmers.

Such a change creates totally new career opportunities in established agriculture. Much of the industrial needs for resins, rubber, plasticizers, coatings, and polymers presently is met with imports because food production generally has been the most profitable alternative for U.S. agriculture. Continued changes and competitive advantages for food crops, combined with new technology for industrial-use products, could result in a large demand base for specialized crops in the future. This development will occur only if there are careers devoted to it.

The USDA has an office of scientists for such crops. This office has singled out four nonfood crops as the most promising for commercial development.

Crambe. This member of the mustard family produces a seed oil that contains a substance used extensively in polyethylene plastics. The U.S. plastics industry presently imports this product as an additive to promote slippage and to eliminate sticking. Easy-open, easy-close bread wraps use polyethylene film containing this product. Specialists working with this crop report that other constituents in crambe oil have potentials in nylon for fabrics and plastic gears. Crambe is a cool season crop and could be grown from Texas to Wyoming, including the upper Midwest. Its products would substitute for rapeseed oil imported from northern and eastern Europe.

Winter Rapseseed. This is the only commercial source of a long chain fatty acid (22-carbon chain) that is used in manufacture of lubricants, nylons, and plastics. Thus, its use is primarily as an industrial polymer. It could be grown across much of the United States as a winter annual.

Guayule ("wy-oo-lee"). This desert shrub is second to the rubber tree in producing natural rubber. Such natural rubber is preferred over synthetic elastomers for elasticity, resilience, tackiness, and the low heat buildup that is indispensable for bus, truck, and airplane tires. New releases of germplasm were made in 1985 that generate higher rubber yields, and bioregulators have shown promise in increasing rubber synthesis as well. Its production would be suited in the southwestern states.

Kenaf. This plant is related to cotton and hollyhock and has potential as a pulp fiber crop. It is a fast-growing plant, going from seed to a plant 10 feet tall in less than three months. Already kenaf is used in the pulp industry for newsprint, to a limited extent. For pulp production, the whole plant is processed. Kenaf and wood pulps, when combined, have synergistic effects in increasing paper strength and quality and reducing ink use in newsprint. Kenaf is a light bulky crop; hence, distant transport is uneconomical. Production from one acre (eight tons) would yield about four tons of kraft pulp. It is best adapted to the lower belt states from the Carolinas to California.

ANIMAL PRODUCTION—PLANT CONVERSION

In the strict sense, animal production is a conversion process in the food chain. Plants remain the basic means of food production, but this conversion of plant material to meat products utilizes plant material otherwise useless to humans.

As with plant production, research and development careers in animal production center on improved efficiency. Nutritionists, geneticists, and other career specialists have developed poultry strains that can convert as little as two pounds of grain into one pound of live weight. In beef, the conversion rate is about five-to-seven, and it is four-to-five for pork.

The basic paths to these efficient conversions are similar to those in the plant sciences—genetics, disease and pest control, nutrition, growth stimulators, and management. Although fundamentally similar, the details are far different. For those with an education in the sciences, there are many career opportunities in the animal industries for research and development. The following examples illustrate the type of careers in this specialized, highly technical segment of agriculture.

Genetics

Genetic engineers and molecular biologists, by unraveling DNA and manipulating genes, may create future "biological factories" involving the oviduct of a chicken or the mammary gland of a cow. Research on pituitary hormones in broilers, for example, may result in producing a pound of broiler growth with only 1.5 pounds of food ration.

Physiologists who specialize in animal reproduction are changing the way livestock reproduce through use of embryo transplants. The transfer of genes in livestock, in addition to conventional genetics, depends to a large extent upon the ability to collect and transfer the fertilized egg, or embryo, nonsurgically. This act is called embryo transfer.

Animal Health

The USDA has reported that animal diseases cut the annual gross farm income from livestock by an estimated 20 percent. In one year, such a loss can reach or surpass 18 million. With such a loss, it is understandable that many commercial companies, especially those with a base of pharmaceutical products, are pushing strongly into veterinary biotechnology.

Research careers in animal health are concentrating on vaccine development, growth hormones, and new biotechnology products called *probes* and *vectors*. Monoclonal antibody diagnosis and treatment is another area of concentrated research.

Gene engineering is employed with vaccine work, with recombinant DNA the principal tool. Such a modified DNA can control reproduction of a virus, and in this way be made incapable of causing a disease. As with plants, enzymes are used in animal genetic engineering as chemical razors to slice certain genetic information from an organism's DNA. This segment is then spliced into the DNA of a bacterial virus (plasmid).

Successful development of vaccines from such biotechnology is expected for foot-and-mouth disease, scours, and coccidiosis.

SUMMARY

The careers in agricultural research and development described in this chapter center on plant and animal production. They are only examples, for there are many other careers in these areas, as well.

In addition, there are just as many careers in areas of food harvesting, storage, processing, and merchandising. Many of these are mentioned in Chapter 4.

There is hardly a field of science or technology that does not offer career opportunities for researchers in agriculture. Biology, in a general sense, is the centerpiece for many of these careers, whether one is seeking to transfer a gene by splicing a molecule of DNA or to develop irradiation techniques for processing healthful foods.

RESEARCH AND DEVELOPMENT OPPORTUNITIES

As a summary, the type of careers in agriculture with opportunities in research and development are represented by the following:

- Agronomist
- Animal cytologist
- Animal geneticist
- Animal nutritionist
- Animal physiologist
- Bacteriologist
- Biochemist
- Bioengineer
- Botanist
- Embryologist
- Entomologist
- Food chemist
- Hydrologist
- Microbiologist
- Nematologist
- Organic chemist
- Parasitologist
- Patent rights specialist
- Pharmaceutical chemist
- Plant cytologist
- Plant nutritionist
- Plant pathologist
- Poultry scientist
- Reproductive physiologist
- Soil scientist
- Veterinary pathologist
- Virologist
- Zoologist

CHAPTER 9

AGRICULTURAL PUBLIC RESEARCH CAREERS

There are many career groups that provide information to all segments in agriculture and are vital to moving products through the marketing system. These support groups add depth to the structure of agriculture and provide much of the thrust for continued advancement in production efficiency, innovation in processing, and improvements in marketing.

As seen elsewhere in this text, there are numerous careers in federal and state government agencies, industry groups, news media, and commodity trading systems. In addition, agencies in the USDA, combined with the network of state universities and state departments of agriculture, offer opportunities for such careers. Many professional areas in these public agencies have counterparts in the private industries. From such pairing of career roles, strong teamwork develops that serves the needs of agriculture better than either group working alone.

PUBLIC RESEARCH

Many of the research career opportunities described in the preceding chapter are in public research agencies. The "team" alignment between researchers in public and private sectors provides not only a supportive relationship but a highly synergistic value where the benefits from the combined efforts are greater than the sum of the two sectors working independently.

Careers in public research primarily focus on solving problems. This orientation for the physical scientist may center on genetic characteristics of wheat varieties for tolerance to aluminum toxicity in acid soils. For the social scientist, it may focus on economics of export enhancement programs in overcoming problems involving demand for U.S. products overseas. Such an orientation for public research careers on problems is somewhat different than that for many research careers in the private sector, which are more closely oriented to product development, even though product development research itself is closely aligned to problems.

Agricultural Research Service (ARS)

The USDA offers many career opportunities in its Agricultural Research Service (ARS). Until recently, this agency had 15 research centers in the United States, with nearly all of its work in the physical and biological sciences. With the trend to downsize government, it is possible that some of these centers may be combined or closed in the future. The following list of research centers illustrates the breadth of research by this single agency:

Human Nutrition Research Center	Beltsville, Md.
U.S. National Arboretum	Washington, D.C.
Eastern Regional Research Center	Philadelphia, Pa.
Human Nutrition Research Center on Aging	Boston, Mass.
Western Regional Research Center	Albany, Calif.
Plant Gene Expression Center	Albany, Calif.
Western Human Nutrition Research Center	Presidio, Calif.
Meat Animal Research Center	Ames, Iowa
National Animal Disease Center	Clay Center, Neb.
Northern Regional Research Center	Peoria, Ill.

Human Nutrition Research Center	Grand Forks, N. Dak.
Richard B. Russell Agricultural Research Center	Athens, Ga.
Children Human Nutrition Research Center	Houston, Tex.
Southern Regional Research Center	New Orleans, La.

Some of these sites have specific, narrow fields of research, such as the Plant Gene Expression Center and the Meat Animal Research Center. Others have broad research programs, such as the Richard B. Russell Center, which has a number of career researchers working on environmental issues.

At these centers highly skilled individuals conduct basic research—such as with gene splicing at the Plant Gene Center—assisted by a number of research assistants and technicians. Career opportunities for the latter include the need for analysts in laboratories, specialists in greenhouse research, field technicians to measure sediment loss under rain simulators, and computer specialists to analyze research data.

Economics Research Service (ERS)

In the field of economics, the USDA offers a number of career opportunities in its Economics Research Service (ERS). Among the many roles available in the ERS are the following:

- Produce economic and other social science information.
- Monitor, analyze, and forecast U.S. and world agricultural production and demand.
- Measure cost of, and returns to, agricultural production marketing.
- Evaluate economic performance of U.S. agricultural production and marketing.
- Estimate effects of government policies and programs on farmers and consumers.

Information on the economic state of agriculture is the end result of such functions. Career ERS researchers and staff specialists, whose backgrounds are primarily in economics and statistical analysis, produce information to assist the general public, the Federal executive branch, and Congress. Some researchers in the ERS are involved in evaluating the outlook for agriculture in the former Soviet Union; others study the effects of various trade policies on U.S. exports; while other specialists produce information on costs of fertilizers and pesticides in producing corn or other crops.

Researchers in the ERS are organized in four program divisions:

1. Agriculture and Rural Economy
2. Agriculture and Trade Analysis
3. Commodity Economics
4. Resources and Technology

This structure gives the ERS unusual strength in commodity analysis, evaluation of global developments affecting agriculture, and current situation/outlook assessments. Serving these four major program divisions is a special data services center that utilizes a number of persons in computer careers.

Cooperative State Research Service (CSRS)

A third agency in the USDA that conducts research is the Cooperative State Research Service (CSRS). This agency handles the "experiment station" funds, also commonly called formula funds, which are allocated to individual states. These federal funds are combined with state resources to finance extensive research programs in each state. Thus, the CSRS is primarily an administrative body in the USDA that facilitates cooperative research with a number of state universities.

The research funded through the CSRS covers essentially the total range of disciplines in agriculture, from basic research in biotechnology to education. In the CSRS, there are career researchers who administer research programs in three primary divisions:

1. Plant and Animal Sciences
2. Natural Resources, Food Science, and Social Sciences
3. Office of Grants and Program Systems (includes the Office of Higher Education)

For a list of universities and colleges in the United States offering curricula in agriculture, see appendix A.

Forest Service (FS)

A fourth agency in the USDA that conducts research is the Forest Service (FS). This agency has eight regional centers where career researchers work on problems associated with forestry. Additionally, the FS has a national Forestry Products Research Laboratory at Madison, Wisconsin.

The following list of nine functional research areas illustrates the variety of forestry research areas involving career employees of the FS:

1. Fire and Atmospheric Sciences
2. Forest Insects and Diseases
3. Forest Inventory and Analysis
4. Renewable Research Economics
5. Timber Management Research
6. Watershed Management
7. Wildlife Range and Fish Habitat
8. Forestry Recreation
9. Forestry Products and Harvesting

International Research Programs

Persons seeking agricultural research careers will also find opportunities in a number of international programs, some public and some private. In the public sector, the Agency for International Development (AID) of the U.S. Department of State is an example. It supports extensive research programs in agriculture.

One research organization that AID helps support is the International Fertilizer Development Center. This center, located in Muscle Shoals, Alabama, has a staff of researchers in agronomy, chemical engineering, economics, and other fields that deal with fertilizer product development and fertilizer use in developing countries. This is just one of many centers that provide career opportunities in research, either in national or international programs.

There are many firms in the private sector that offer opportunities in agricultural research. Some of these specialize in toxicology, some in biotechnology, and so forth. Others, such as Wharton Econometrics, specialize in economics and market research at both national and international levels.

CAREER OPPORTUNITIES

There is a wide range of career opportunities in research organizations, both public and private, that support agriculture directly or indirectly through complimentary research conducted by agri-industries themselves. Careers that have not been mentioned in the other chapters include:

- Data processing specialist
- Economist
- Market research economist
- Research administrator
- Statistician

EDUCATION

The demands of agricultural colleges to matriculate more than 25,000 college graduates annually for U.S. agriculture offers many careers in teaching. Many of these careers are at the state universities. There is also a large number of opportunities for educators at high schools with vocational agriculture programs. Additionally, continuing adult-education programs provide other opportunities.

Interests in agriculture often develop during high school; for others, this interest may not develop until after college. Today, nearly half of the freshmen enrolling in college agricultural curricula are from nonfarm backgrounds. Thus, interests developed in early experiences or in learning of life processes, resource management, and other fields often lay the foundation for an agricultural career.

The most extensive program for agriculture at the high school level is the vocational agriculture curriculum provided with initial funding by the National Vocational Education Acts. This program in some 7200 U.S. high schools is usually paired with chapters of the Future Farmers of America (FFA).

There are approximately 12,000 teachers in this high school vocational program. Most of these educators hold degrees in agricultural education or similar major studies in college. In addition to teaching in high school, many vocational agricultural teachers conduct educational programs for adults.

Another large national program that offers careers in education is the federal and state agricultural extension service. This program centers on transfer of technology from researchers to farmers and ranchers. It often is called the cooperative agricultural extension service because of the joint funding from federal, state, and county governments.

Careers in the extension service extend from positions as county extension agents to those of program administrators in the

USDA. Thousands of careers are included in the extension personnel ranks, representing all the disciplines applying to agricultural production, marketing, policy, and communication. At the state land-grant universities, extension specialists are a major part of all agricultural departments, such as agronomy and plant pathology, that have programs oriented to problems of agriculture.

CHAPTER 10

TELLING THE AGRICULTURE STORY

During the late 1800s and early 1900s, farm publications provided the prime source of agri-journalism careers. Today, communication careers span a media spectrum stretching from satellite teleconferencing and instant news event coverage, to television, radio, print, computer graphics, and the use of the Internet.

OPPORTUNITIES IN JOURNALISM

Newspapers need writers with a broad range of agricultural knowledge to translate for readers the "why" of farm programs, or how farm economic developments will affect consumer prices, export demand, world-wide food supplies, or personal nutrition. Every day there is a demand for agricultural news, and it depends on the abilities of agricultural journalists to be abreast of developments and able to write or broadcast these developments—and to interpret them. This is especially true as the question of farm subsidies comes to the fore.

Farm magazines need those skilled in writing, magazine layout and graphics, and photography. Journalists with extended knowledge in agricultural "discipline" or subject matter areas are needed to write and edit for specialized "vertical" publications—or to handle special sections—on livestock, poultry, agri-economics, soil fertility, international issues, home economics, and computer information, among other interests.

Today's farm publications often are highly tailored by farming region as to crop, livestock, business, and consumer interests. Many such magazines restrict subscription eligibility to those who need specialized information, or are sent to readers on a "controlled circulation" basis.

Radio and Television

Many wide-coverage radio and television stations still have full-time farm broadcasters but not as many, perhaps, as a few years ago. Increasingly, nonfarm programming, which may generate higher advertising revenue based on audience numbers, has altered the traditional programming for many stations.

To fill the breach, a phenomenon called "agricultural broadcast networks" has emerged. These networks, in some cases, are as highly specialized in information as the vertical publications mentioned earlier, in that farm news coverage is devoted mainly to one crop, such as tobacco or soybeans. Others cover a range of topics in farming and agribusiness and provide a "magazine" format for radio and TV.

Specialized Communications

The broad scope of agriculture currently involves more than 20 percent of America's working population—not just farmers, of course but farm suppliers, food processors and packagers, scientists, and product development specialists, among many others. To reach this large segment of the nation, plus the consumer as well, career communicators must be skilled in advertising, product marketing, and promotion and public relations.

Agricultural communications careers in advertising include a number of activities. In an advertising agency, for example, there are those who specialize in consumer attitude and preference surveys; advertising copy-writing; design and layout; client and media relations; media placement of advertising; market promotion and planning; and development of sales aids.

Within farm publications and broadcasting staffs, there are advertising sales specialists who work closely with the agency or advertiser in positioning messages that inform readers, listeners, and viewers of available products and services. Directors of public relations, media relations—or, most often, communications relations—in agribusiness and university and government positions often find that they are the media spokespersons for the entity that they represent.

Agricultural Education Careers

"Teach the teacher, train the trainer." Such is the career objective of many communicators in education.

News media tools and technology have been adapted to the classroom, seminar auditorium, and workshop arena. This broadening of careers tied to communications has opened new opportunities for broadcasters, computer communicators, visual specialists, and writers. Three relatively new areas are:

Interactive video, a combination of computer technology and video graphics that utilizes a range of visual, written, and computer programming talents and is a method commanding increasing attention in education.

Teleconferencing, linking a number of individuals by audio or audiovisual satellite transmission, who may be hundreds or thousands of miles apart, into a cohesive problem-solving or discussion group.

Multimedia productions, utilizing a blend of color slides, video tapes, audio sound tracks, and motion picture footage to create a spectacular array of communication techniques.

Those in agricultural education careers at land-grant universities, as well as within agricultural extension work and commercial services, make extensive use of communication techniques. At universities, support staffs of agricultural editors' offices work with researchers, educators, county extension workers, and vocational agricultural and nonfarm news media in the role of media liaison and as information sources for their institutions.

At the county level, extension staffs in agricultural centers may have positions for those highly competent in writing, broadcasting, photography, visual-aid production, multimedia presentation, or other communication areas and skills.

Other Opportunities

Many international opportunities await the proven agricultural communicator. Numerous colleges and universities, benevolent and religious services, major foundations, and state and federal agencies seek trained agricultural communication specialists to work outside the U.S., primarily in developing nations.

Again, with the growing use of computers and electronic message capabilities, a number of specialized information services to agriculture have emerged. Most provide farmers and other agriculturists with near-instant updates on stock and bond markets, livestock and grain prices, government announcements, and political and business news.

EDUCATIONAL OPPORTUNITIES

A number of universities offer bachelor degrees in agricultural communications or journalism. Several offer specialized work leading to graduate degrees, as well.

Usually, the student interested in such careers follows one of two collegiate routes. One consists of coursework leading to the communications or journalism degree. The other may be a selective program allowing one to major in an agricultural discipline, such as economics or soil chemistry, and maintain a minor study in journalism areas of choice or major in journalism with a broad range of selected agricultural courses.

Some of the colleges and universities which traditionally have offered educational opportunities in agricultural communications or journalism follows:

Abraham Baldwin Agriculture College
Tifton, GA 31794

Auburn University
Auburn, AL 36830

California State University
San Luis Obispo, CA 93401

Colorado State University
Fort Collins, CO 80521

Cornell University
Ithaca, NY 14850

Iowa State University
Ames, IA 50010

Kansas State University
Manhattan, KS 66502

Michigan State University
East Lansing, MI 48823

Mississippi State University
Mississippi State, MS 39762

New Mexico State University
University Park, NM 88001

North Carolina State University
Raleigh, NC 27607

Ohio State University
Columbus, OH 43210

Oklahoma State University
Stillwater, OK 74074

Oregon State University
Corvallis, OR 97331

Pennsylvania State University
University Park, PA 16802

Purdue University
West Lafayette, IN 47907

South Dakota State University
Brookings, SD 57006

Texas A & M University
College Station, TX 77843

Texas Tech University
Lubbock, TX 79401

University of Arkansas
Fayetteville, AR 72701

University of Arizona
Tucson, AZ 85721

University of Delaware
Newark, DE 19711

University of Florida
Gainesville, FL 32601

University of Georgia
Athens, GA 30601

University of Idaho
Moscow, ID 83840

University of Illinois
Urbana, IL 61801

University of Kentucky
Lexington, KY 40506

University of Maine
Orono, ME 04473

University of Minnesota
St. Paul, MN 55108

University of Minnesota-Crookston
Crookston, MN 56716

University of Missouri
Columbia, MO 65201

University of Nebraska
Lincoln, NE 68503

University of Rhode Island
Kingston, RI 02881

University of Wisconsin
Madison, WI 53706

University of Wyoming
Laramie, WY 82070

Utah State University
Logan, UT 84321

Washington State University
Pullman, WA 99163

AGRICULTURAL PROFESSIONAL ASSOCIATIONS

In addition to selected universities, a number of professional associations maintain active programs promoting agricultural communications, and some offer scholarships and intern opportunities. A list of the major ones follows:

Agricultural Communicators in Education (college and government)
University of Florida
Building 116, 6011 FAS
Gainesville, FL 32611

Agricultural Relations Council (public relations/advertising)
1629 K Street, N.W.
Washington, DC 20006

American Agricultural Editors' Association (farm magazines)
612 West 22 Street
Austin, TX 78705-5116

Cooperative Communicators Association (editorial staffs at cooperatives)
2263 East Bancroft
Springfield, MO 65804

National Agri-Marketing Association (marketing and advertising)
P.O. Box 7912
Overland Park, KS 66207

National Association of Agricultural Journalists (daily and weekly newspapers)
312 Valley View
Huron, OH 44839

National Association of Farm Broadcasters (radio and TV)
26 East Exchange Street
St. Paul, MN 55101.

You may also want to check the "Career Curricula" in the front of *Lovejoy's College Guide,* published annually by Prentice-Hall.

CAREER OPPORTUNITIES

A listing of some of the specialty areas in agricultural communications would include: advertising sales; advertising writer; program specialist; agricultural marketing specialist; agricultural movie-video producer; agricultural photographer; agricultural reporter or columnist for a metropolitan newspaper or news-wire service; broadcast news specialist for local/network news; college agricultural communicator/editor/broadcaster; communicator with state/federal agencies; computer communications specialist; farm editor, rural newspaper; farm magazine editor; farm radio and television editor; food and nutrition, home economics writer/editor; graphics artist; public relations specialist; and telecommunications specialist.

CHAPTER 11

PREPARING FOR YOUR AGRICULTURAL CAREER

Do you know what you really want to do? Have you ever tried writing down your life objective? When we have to turn ideas into words we are forced to do some real thinking. Every man and woman should have a Big Idea, a purpose or goal. Some young people know what they want to do sooner than others. They can start preparing for their careers by selecting the high-school or college courses which will best prepare them for their college or perhaps graduate work.

Don't despair if you haven't decided on your Big Idea. You will sooner or later, but you can hasten that time by giving the matter a lot of thought, reading about various careers, and talking with teachers, guidance counselors, and others who can help you. Be prepared for the chance that you may later change your mind. It has been estimated that early in the next century workers will change professions on an average of three times and jobs six times. Should you now decide against entering some branch of agriculture as a career, it is never too late to reverse that decision and get started in this field.

One way to learn more about farming and agriculture is to find an internship, a summertime chance to find out whether or not a particular job area is for you. Many such paid and non-paid opportunities exist for high school and college students; you have to find the right one for you. Ask your guidance counselor, state employment service, friends, and business acquaintances for suggestions. Visit the local bookstore and public library to see what books they may have, and consult the Petersons Guide on *Internships.*

The man or woman who is content to drive a tractor all his or her life has a job. However, the individual who drives the tractor because he or she wants to start at the bottom of the agricultural field and learn everything possible about the business, so as to eventually advance to a top-management position, has begun a career.

After the turn of this century, a high-school diploma plus at least one year of vocational or technical training will be the minimum education needed to enter any worthwhile position. Training at a two-year technical school or community college may not be as expensive as you think. College costs vary but you do not have to attend an Ivy League school or one of the "Big Ten." There are hundreds of state universities and small, less well-known colleges where you can obtain excellent undergraduate training.

Many students earn part of their expenses by waiting on tables, working in the library, having a laundry route, typing, tutoring, or finding a part-time job. Summer employment also can earn dollars for your education. Of course, scholarship aid and student loans are also available.

If you are determined to obtain a college or postgraduate education, nothing need stand in your way. Even if you cannot enter college immediately after graduation, remember it is always possible to attend at a later time. Once you have obtained your undergraduate college training, you will find it easier to finance your graduate work.

You may want to investigate the educational and civilian employment possibilities and opportunities available to those who have served in the Armed Forces following graduation from high school or college.

You too can follow in the path of the countless men and women who have found meaning and success in an agricultural career. Decide what you want to do, get your training, and experience a lifetime of satisfaction and a sense of purpose.

APPENDIX A

UNIVERSITIES AND COLLEGES IN THE UNITED STATES OFFERING CURRICULA IN AGRICULTURE

The universities in the following list can be useful in providing information on various agricultural curricula, in addition to information on research careers. Most of the universities in the list work with the CSRS in joint research efforts.

Alabama

Alabama A&M University
Normal, AL 35762

Auburn University
Auburn University, AL 36849

Tuskegee Institute
Tuskegee Institute, AL 36088

Alaska

University of Alaska
Agricultural Experiment Station
Palmer, AK 99645

Arizona

Arizona State University
Tempe, AZ 85281

University of Arizona
Tucson, AZ 85721

Arkansas

Southern Arkansas University
Magnolia, AR 71753

University of Arkansas
Fayetteville, AR 72701

University of Arkansas
Monticello, AR 71655

Source: American Society of Agronomy, 677 South Segoe Road, Madison, WI 53711

University of Arkansas
Pine Bluff, AR 71601

California

California Polytechnic State University
San Luis Obispo, CA 93407

California State Polytechnic University
Pomona, CA 91768

California State University
Chico, CA 95926

California State University
Fresno, CA 93740

Humboldt State University
Arcata, CA 95521

University of California
Davis, CA 95616

University of California
Riverside, CA 92521

Colorado

Colorado State University
Fort Collins, CO 80523

Fort Lewis College
Durango, CO 81301

Connecticut

University of Connecticut
Storrs, CT 06268

Delaware

Delaware State College
Dover, DE 19901

University of Delaware
Newark, DE 19711

Florida

Florida Southern College
Lakeland, FL 33802

University of Florida
Gainesville, FL 32611

Georgia

Abraham Baldwin Agricultural College
Tifton, GA 31794

Berry College
Mount Berry, GA 30149

Fort Valley State College
Fort Valley, GA 31030

University of Georgia
Athens, GA 30602

Hawaii

University of Hawaii
Honolulu, HI 96822

Idaho

College of South Idaho
Twin Falls, ID 83301

Ricks College
Rexburg, ID 83440

University of Idaho
Moscow, ID 83843

Illinois

Illinois State University
Normal, IL 61761

Southern Illinois University
Carbondale, IL 62901

University of Illinois
Urbana, IL 61801

Western Illinois University
Macomb, IL 61455

Indiana

Purdue University
West Lafayette, IN 47907

Iowa

Iowa State University
Ames, IA 50011

Kansas

Fort Hays State University
Hays, KS 67601

Kansas State University
Manhattan, KS 66506

McPherson College
McPherson, KS 67460

Kentucky

Morehead State University
Morehead, KY 40351

Murray State University
Murray, KY 42071

University of Kentucky
Lexington, KY 40506

Western Kentucky University
Bowling Green, KY 42101

Louisiana

Louisiana State University
Baton Rouge, LA 70803

Louisiana Tech University
Ruston, LA 71272

McNeese State University
Lake Charles, LA 70609

Nichols State University
Thibodaux, LA 70310

Northeast Louisiana University
Monroe, LA 70209

Northwestern State University
Natchitoches, LA 71497

Southeastern Louisiana University
Hammond, LA 70402

Southern University
Baton Rouge, LA 70813

University of Southwestern Louisiana
Lafayette, LA 70504

Maine

University of Maine
Orono, ME 04469

Maryland

University of Maryland
College Park, MD 20742

University of Maryland
Princess Anne, MD 21853

Massachusetts

University of Massachusetts
Amherst, MA 01003

Michigan

Michigan State University
East Lansing, MI 48824

Michigan Technological University
Hancock, MI 49930

Northern Michigan University
Marquette, MI 49855

Minnesota

University of Minnesota
St. Paul, MN 55108

University of Minnesota
Technical College
Crookston, MN 56716

University of Minnesota
Technical College
Waseca, MN 56093

Mississippi

Alcorn State University
Lorman, MS 39096

Mississippi State University
Mississippi State, MS 39762

Missouri

Central Missouri State
University
Warrensburg, MO 64093

Lincoln University
Jefferson City, MO 65101

Missouri Western State College
St. Joseph, MO 64507

Northeast Missouri State
University
Kirksville, MO 63501

Northwest Missouri State
University
Maryville, MO 64468

Southwest Missouri State
University
Springfield, MO 65802

University of Missouri
Columbia, MO 65211

Montana

Montana State University
Bozeman, MT 59717

Northern Montana College
Havre, MT 59501

Nebraska

University of Nebraska
Lincoln, NE 68583

Nevada

University of Nevada
Reno, NV 89557

New Hampshire

University of New Hampshire
Durham, NH 03824

New Jersey

Rutgers University
New Brunswick, NJ 08903

New Mexico

New Mexico State University
Las Cruces, NM 88003

New York

Cornell University
Ithaca, NY 14853

North Carolina

North Carolina A&T State
University
Greensboro, NC 27411

North Carolina State University
Raleigh, NC 27650

North Dakota

North Dakota State University
Fargo, ND 58105

Ohio

Ohio State University
Columbus, OH 43210

Wilmington College
Wilmington, OH 45177

Oklahoma

Cameron University
Lawton, OK 73505

Langston University
Langston, OK 73050

Oklahoma Panhandle State University
Goodwell, OK 73939

Oklahoma State University
Stillwater, OK 74074

Oregon

Oregon State University
Corvallis, OR 97331

Pennsylvania

Delaware Valley College of Science and Agriculture
Doylestown, PA 18901

Pennsylvania State University
University Park, PA 16802

Temple University
Ambler, PA 19002

Puerto Rico

University of Puerto Rico
Mayaguez, PR 00708

University of Puerto Rico
Rio Piedras, PR 00928

Rhode Island

University of Rhode Island
Kingston, RI 02881

South Carolina

Clemson University
Clemson, SC 29631

South Dakota

South Dakota State University
Brookings, SD 57007

Tennessee

Austin Peay State University
Clarksville, TN 37040

Middle Tennessee State University
Murfreesboro, TN 37130

Tennessee State University
Nashville, TN 37208

Tennessee Technological University
Cookeville, TN 38505

University of Tennessee
Knoxville, TN 37901

University of Tennessee
Martin, TN 38238

Texas

Abilene Christian University
Abilene, TX 79601

East Texas State University
Commerce, TX 75428

Prairie View A&M University
Prairie View, TX 77445

Sam Houston State University
Huntsville, TX 77341

Southwest Texas State University
San Marcos, TX 78666

Stephen F. Austin State University
Nacogdoches, TX 75962

Texas A&I University
Kingsville, TX 78363

Texas A&M University
College Station, TX 77843

Texas Tech University
Lubbock, TX 79409

West Texas State University
Canyon, TX 79015

Utah

Brigham Young University
Provo, UT 84602

Utah State University
Logan, UT 84322

Vermont

University of Vermont
Burlington, VT 05405

Virginia

Old Dominion University
Norfolk, VA 23508

Virginia Polytechnic Institute and State University
Blacksburg, VA 24061

Virginia State College
Petersburg, VA 23806

Washington

University of Washington
Seattle, WA 98195

Washington State University
Pullman, WA 99164

West Virginia

West Virginia University
Morgantown, WV 26506

Wisconsin

University of Wisconsin
Green Bay, WI 54302

University of Wisconsin
Madison, WI 53706

University of Wisconsin
Platteville, WI 53818

University of Wisconsin
River Falls, WI 54022

University of Wisconsin
Stevens Point, WI 54481

Wyoming

University of Wyoming
Laramie, WY 82071

APPENDIX B

CAREER STATISTICS FOR SELECTED AGRICULTURAL JOBS

Almost all of the statistics presented below were compiled by the U.S. Bureau of Labor Statistics in 1992. Because, as we pointed out in the text, the agricultural field embraces such a wide variety of careers, only the most pertinent jobs in the support, subprofessional, and professional categories are included below to give some idea of earnings possibilities. Most of the figures are in the medium range, based on high and low salaries, but a few give a range which reflects education and experience.

The capital letters refer to the job outlook as defined below:

A. Will grow much faster than the average.
B. Will grow faster than the average.
C. Will grow about as fast as the average.
D. Will grow more slowly than the average.
E. Will show little change—increase or decrease.

The final numeral refers to the minimum education required as defined below:

1. High school diploma.
2. High school plus one year in a vocational/technical school.
3. High school plus two years in a vocational/technical school.
4. Four years college.
5. Undergraduate degree plus one year graduate school.

6. Undergraduate degree plus six to nine years graduate school, according to program selected.

Career	Outlook/Salary/Education
Administrative Services Manager	D; $40,000; 2
Agricultural Scientists	C; Beginning salaries for B.S. in animal science averaged $20,189 and $22,150 for plant science; average Federal salaries $55,631 for animal science, $45,911 for agronomy, $43,033 for soil science, $44,492 for horticulture, $53,889 for entomology; 4
Billing Clerks	E; $18,400; 1
Biological and Life Scientists	B; $34,500; 4
Bookkeeping, Accounting, and Auditing Clerks	E; $19,000; 1
Chemists	C; $24,000; 4
College and University Faculty	C; average for instructors, $27,700; 4
Employment Interviewers	C; $20,000; 4
Engineers	C; $31,000; 4
Farm Managers	E; Highest over $696 per week, lowest under $185 per week, 4
File Clerks	C; $15,700; 1
Foresters and Conservation Specialists	D; $18,340–22,717 (1993) starting salaries; 4
Forestry Aides (Federal Government employees)	E; $26,600 average; 3
Forestry and Logging Workers	E; $159–556+ per week; 1
Gardeners and Groundskeepers	A; $275 per week; 1
General Office Clerks	C; $18,500; 1
Maintenance Mechanics	B; $9.37 per hour; 1
Marketing, Advertising, and Public Relations Managers	A; less than $22,000–$79,000+; 4

Career	Outlook/Salary/Education
Payroll Clerks	E; $21,000; 2
Personnel, Training, and Labor Relations Specialists and Managers	B; $32,000; 4
Purchasers and Buyers	D; $33,067; 4
Public Relations Specialists	C; $32,000; 4
Receptionists	A; $400 per week; 1
Secretaries	D; $26,700; 2
Stenographers	E; $410 per week, 2
Telephone Operators	E; $385 per week, 1
Typists, Word Processors, and Data Entry Keyers	E; $20,000; 1

APPENDIX C

SUGGESTED READINGS

The following books published by National Textbook Company provide excellent career background material for many of the job areas covered in this book. Should you need to obtain additional information on any of the subjects mentioned in the text, consult your public library catalog and the subject volumes of *Books in Print.* If you want titles unobtainable in your local library, the librarian may be able to obtain them through interlibrary loans.

SUPPORT SECTOR

Arpan, Jeffrey S. *Opportunities in International Business Careers.*

Banning, Kent. *Opportunities in Purchasing Careers.*

Basye, Anne. *Opportunities in Direct Marketing Careers.*

Dahm, Ralph, and James Brescoll. *Opportunities in Sales Careers.*

Dolber, Roslyn. *Opportunities in Retailing Careers.*

Ettinger, Blanche. *Opportunities in Office Occupations Careers.*

———. *Opportunities in Secretarial Careers.*

Gaylord, Gloria, and Glenda E. Ried. *Careers in Accounting.*

Heim, Kathleen, and Margaret Myers. *Opportunities in Library and Information Careers.*

Kanter, Elliott. *Opportunities in Computer Maintenance Careers.*

Kling, Julie Lepick. *Opportunities in Computer Science Careers.*

Noerper, Norman. *Opportunities in Data Processing Careers.*

Paradis, Adrian A. *Opportunities in Cleaning Services Careers.*
———. *Opportunities in Vocational and Technical Careers.*
Pattis, S. William. *Careers in Advertising.*
Place, Irene. *Opportunities in Business Management Careers.*
Ring, Trudy. *Careers in Finance.*
Rosenberg, Martin. *Opportunities in Accounting Careers.*
Stair, Lila B. *Careers in Computers.*
———. *Careers in Marketing.*
Stair, Lila B., and Dorothy Domkowski. *Careers in Business.*
Steinberg, Margery. *Opportunities in Marketing Careers.*

COLLEGE PREPARATION

Breenan, Moya, and Sarah Briggs. *How to Apply to American Colleges.*
Eberts, Marjorie, and Margaret Gisler. *How to Prepare for College.*
Jackson, Acy L. *How to Prepare Your Curriculum Vitae.*
Paradis, Adrian A. *Opportunities in Part-Time and Summer Jobs.*
Rubinfeld, William A. *Planning Your College Education.*
Schrank, Louise Welsh. *How to Choose the Right Career.*

There are also innumerable college guides, directories of scholarships, and guides to scholarships and loans. Because few libraries can afford to buy all of them, you will have to use those available to you. Titles of some of the best known include: *Lovejoy's College Guide, The Scholarship Guide, Directory of Financial Aids for Women, Graduate Scholarship Directory, Free Money for Graduate School.* In addition there are the following Petersons Guides: *Four-Year Colleges, Two-Year Colleges, Paying Less for College, Graduate and Professional Programs, Internships.*

JOURNALISM AND PUBLIC RELATIONS

Bone, Jan. *Opportunities in Telecommunications Careers.*

Basye, Anne. *Opportunities in Telemarketing Careers.*

Carter, Robert A. *Opportunities in Book Publishing Careers.*

Ellis, Elmo I. *Opportunities in Broadcasting Careers.*

Ferguson, Donald L., and Jim Patten. *Opportunities in Journalism Careers.*

Foote-Smith, Elizabeth. *Opportunities in Writing Careers.*

Goldberg, Jan. *Careers in Journalism.*

Gould, Jay, and Wayne Losano. *Opportunities in Technical Communications Careers.*

———. *Opportunities in Technical Writing and Communications Careers.*

Noronha, Shonan F. R. *Careers in Communications.*

———. *Opportunities in Television and Video.*

Pattis, S. William. *Opportunities in Advertising Careers.*

———. *Opportunities in Magazine Publishing Careers.*

Rotman, Morris B. *Opportunities in Public Relations Careers.*

Tebbel, John, *Opportunities in Newspaper Publishing Careers.*

FARMING AND AGRICULTURE

Arpan, Jeffrey. *Opportunities in International Business Careers.*

Dahm, Ralph, and James Brescoll. *Opportunities in Sales Careers.*

Dolber, Roslyn. *Opportunities in Retailing Careers.*

Garner, Geraldine O. *Opportunities in Engineering Careers.*

Goldberg, Jan. *Opportunities in Horticulture Careers.*

Hagerty, D. Joseph, and Louis F. Cahn. *Opportunities in Civil Engineering Careers.*

Hagerty, D. Joseph, and John E. Herr, Jr. *Opportunities in Engineering Technology Careers.*

Kinney, Jane, and Mike Fasulo. *Careers for Environmental Types and Others Who Respect the Earth.*

Lee, Mary Price, and Richard S. *Opportunities in Animal and Pet Care.*

Miller, Louise. *Careers for Animal Lovers and Other Zoological Types.*
———. *Careers for Nature Lovers and Other Outdoor Types.*
———. VGM's Career Portraits: Animals
Paradis, Adrian A. *Opportunities in Transportation Careers.*
Rowh, Mark. *Opportunities in Waste Management Careers.*
Stair, Lila B. *Careers in Marketing.*
Steinberg, Margery. *Opportunities in Marketing Careers.*
Swope, Robert E., and Sarah Nikesell. *Opportunities in Veterinary Medicine.*
Wille, Christopher M., *Opportunities in Forestry Careers.*
Winter, Charles A. *Opportunities in Biological Science Careers.*
Woodburn, John R. *Opportunities in Chemistry Careers.*
———. *Opportunities in Energy Careers.*

APPENDIX D

ORGANIZATIONS TO CONTACT FOR FURTHER INFORMATION

JOURNALISM, PUBLISHING, RADIO, & TELEVISION

American Federation of Television and Radio Artists
1350 Avenue of the Americas
New York, NY 10019

Association of American Publishers
220 East 23 Street
New York, NY 10010

Broadcast Education Association
1771 N Street, N.W.
Washington, DC 20006

National Academy of Television Arts and Sciences
111 West 57 Street
New York, NY 10019

National Association of Broadcasters
1771 N Street, N.W.
Washington, DC 20006

National Newspaper Association
1627 K Street, N.W.
Washington, DC 20006
(A pamphlet, "A Career in Newspapers," is obtainable on request.)

Public Relations Society of America and the Public Relations Student Society of America
33 Irving Place
New York, NY 10003
(A directory of schools offering degree programs or sequence of study in public relations, and a brochure on public relations available for $10 and $2 respectively.)

Note: *Editor and Publisher International Year Book,* available in most public libraries, gives names and locations of newspapers and schools with departments of journalism.

FARMING AND AGRICULTURE

Agency for International Development
Office of Internal Affairs
Washington, DC 20523-0001

Agriculture and Agri-Food Canada
930 Carling Avenue
Ottawa, Ontario, Canada K1A 0C7

Agricultural Canada, Communications Branch
930 Carling Avenue
Ottawa, Ontario, Canada K1A 0C7

Agricultural Marketing Service
U.S. Department of Agriculture
P.O. Box 96456
Washington, DC 20250

Agricultural Research Service
U.S. Department of Agriculture
Beltsville, MD, 20705

Agricultural Service
U.S. Department of Agriculture
6303 Ivy Lane
Greenbelt, MD 20770

Agricultural Stabilization and Conservation Service
U.S. Department of Agriculture
P.O. Box 2415
Washington, DC 20013

American Association of Meat Processors
P.O. Box 269
Elizabethtown, PA 17022

American Chemical Society
1155 16 Street, N.W.
Washington, DC 20036

American Dairy Association
10255 West Higgins Road
Rosemont, IL 60018-5616

American Farm Bureau Federation
225 Touhy Avenue
Park Ridge, IL 60068

American Fiber, Textile, Apparel Coalition
1801 K Street, N.W.
Washington, DC 20006

American Forest and Paper Association
1250 Connecticut Avenue, N.W.
Washington, DC 20036

American Forestry Association
1516 P Street, N.W.
Washington, DC 20015

American Institute of Chemical Engineers
345 East 47 Street
New York, NY 10017

American Management Association
135 West 50 Street
New York, NY 10020

American Meat Institute
P.O. Box 3556
Washington, DC 20007

American Seed Trade Association
601 13 Street, N.W.
Washington, DC 20005

American Sheep Industry Association
69115 Yosemite
Englewood, CO 80112-1414

American Society of Agronomy, Crop Science Society of America, Soil Science Society of America
677 South Segoe Road
Madison, WI 53711

American Society of Farm Managers and Rural Appraisers
950 South Cherry Street
Denver, CO 80222

American Soybean Association
P.O. Box 419200
St. Louis, MO 63141

Animal and Plant Health Service
U.S. Department of Agriculture
Washington, DC 20250

Board of Trade of the City of Chicago
141 West Jackson Boulevard
Chicago, IL 60604

Canadian Dairy Commission
1525 Carling Avenue
Ottawa, Ontario, Canada K1A 0Z2

Canadian Grain Commission
303 Main Street
Winnepeg, MB, Canada R3C 3G8

Canadian International Trade Tribunal
333 Laurier Avenue
Ottawa, Ontario, Canada K1A 0G7

Canadian Wheat Board
423 Main Street
Winnipeg, MB, Canada R3C 2P5

Economic Research Service
U.S. Department of Agriculture
Washington, DC 20005-4789

Economics Management Staff, same address as above

Employment and Immigration Canada
140 Promenade du Portage
Ottawa-Hull, Ottawa, Canada K1A 0J9

Environment Canada
10 Wellington Street
Hull, PQ, Canada K1A 0H3

Environmental Protection Agency
401 E Street, S.W.
Washington, DC 20460

Farm Credit Corporation Canada
1800 Hamilton Street
Regina, SA Canada S4P 4L3

Farm Equipment Manufacturers Association
243 North Lindbergh Boulevard
St. Louis, MO 63141

Federal Grain Inspection Service
U.S. Department of Agriculture
Washington, DC 20250

Fertilizer Institute
501 2 Street, N.E.
Washington, DC 20002

Food and Agricultural Careers for Tomorrow
Purdue University
1140 Agricultural Administration Building
West Lafayette, IN 47907-1140

Food Marketing Institute
800 Connecticut Avenue, N.W.
Washington, DC 20005

Food Processing Machinery and Supplies Association
200 Daingerfield Drive
Alexandria, VA 22314

Food Processors Institute
1401 New York Avenue N.W.
Washington, DC 20005

Forest Service
U.S. Department of Agriculture
P.O. Box 916090
Washington, DC 20090-6090

Food Safety and Inspection Service
U.S. Department of Agriculture
Washington, DC 20250

Foreign Agriculture Service, same address as above.

Future Farmers of America, see National FFA Organization

Genetics Society of America
9650 Rockville Pike
Bethesda, MD 20814-3995
(A brochure, "Solving the Puzzle: Careers in Genetics," free on request.)

Grain Elevator and Processing Society
P.O. Box 15026
Minneapolis, MN 55415

Higher Education Programs Office
U.S. Department of Agriculture
Washington, DC 20250

Institute of Food Technologists
221 North LaSalle Street
Chicago, IL 60601

Livestock Conservation Institute
6414 Copps Avenue
Madison, WI 53716

Marketing Research Association
2189 Silas Deane Highway
Rocky Hill, CT 06067

National Agriculture Chemicals Association
1155 15 Street, N.W.
Washington, DC 20005

National Association of Business Economists
28790 Chagrin Road
Cleveland, OH 44122

National Association of Conservation Districts
509 Capitol Court, N.E.
Washington, DC 20002

National Cotton Council of America
P.O. Box 12285
Memphis, TN 38182-0285

National FFA Organization
P.O. Box 15160
Alexandria, VA 22309-0160 (Formerly Future Farmers of America.)

National Food Processors Association
1401 New York Avenue, N.W.
Washington, DC 20005

National Frozen Food Association
P.O. Box 6069
Harrisburg, PA 17112

National Pest Control Association
8100 Oak Street
Dunn Loring, VA 22027

National Retail Federation
701 Pennsylvania Avenue, N.W.
Washington, DC 20004

National Science Foundation
1800 G Street, N.W.
Washington, DC 20550

National Young Farmers Educational Association
P.O. Box 223
Sheridan, IN 46069

New York Commodity Exchange
4 World Trade Center
New York, NY 10048

Packaging Education Foundation
481 Carlisle Drive
Herndon, VA 22070

Packaging Machinery Manufacturers Institute
1343 L Street, N.W.
Washington, DC 20005

Peace Corps
Office of Volunteer Services
Washington, DC 20526

Potash and Phosphate Institute
655 Engineering Drive
Norcross, GA 30092

Poultrymen's Cooperative Association
1863 Service Court
Riverside, CA 92507

Professional Farmers of America
P.O. Box 6
Cedar Falls, IA 50613

Public Health Service
200 Independence Avenue, S.W.
Washington, DC 20201

Soil Conservation Service
U.S. Department of Agriculture
P.O. Box 2890
Washington, DC 20013

Statistics Canada
R. H. Coats Building, Tunney's Pasture
Ottawa, Ontario, Canada K1A 0T6

United Engineering Trustees
345 East 47 Street
New York, NY 10017

U.S. Department of Agriculture
Washington, DC 20250

U.S. Geological Survey
119 National Center
Reston, VA 22092

U.S. Trade Representative
600 17 Street, N.W.
Washington, DC 20506

United States Wheat Association
1620 Eye Street, N.W.
Washington, DC 20006

Water Quality Association
4151 Neperville Road
Lisle, IL 60532

Note: For names and addresses of other Canadian government bureaus, consult the *Canadian Almanac and Directory,* published annually and available at many libraries.